# International Civil Engineering and Architecture Conference (CEAC 2021)



Held online due to COVID-19

IOP Conference Series: Materials Science and Engineering
Volume 1165

Singapore
11 - 14 March 2021

ISBN: 978-1-7138-3430-4
ISSN: 1757-8981

**Printed from e-media with permission by:**

Curran Associates, Inc.
57 Morehouse Lane
Red Hook, NY 12571

**Some format issues inherent in the e-media version may also appear in this print version.**



**Additional copies of this publication are available from:**

Curran Associates, Inc.
57 Morehouse Lane
Red Hook, NY 12571 USA
Phone: 845-758-0400
Fax: 845-758-2633
Email: curran@proceedings.com
Web: www.proceedings.com

# TABLE OF CONTENTS

IOP Conf. Series: Materials Science and Engineering 1165 (2021) 011001 doi:10.1088/1757-899X/1165/1/011001

# Preface

2021 International Civil Engineering and Architecture Conference (CEAC 2021) was held online during March 11-14, 2021. Due to the evolving public health concerns and travel advisories issued for the avoidance of non-essential travel in view of COVID-19, the conference CEAC 2021 has been held virtually hosted on Zoom video conference platform. It was a fully synchronized online conference that had live Q&A, participants jointly listened to presentations, and attended other live events associated with the conference.

Scientists from China, Indonesia, Thailand, Qatar, Pakistan, Japan, and so on took part in the scientific forum. And for plenary speech, each plenary speech lasted for 40 minutes, including 5 minutes for Q&A while for other sessions, there was 15 minutes for each presentation, including 2 minutes for Q&A. The interaction session was real experience to the participants and they had good exposure with foreign researchers, students and other delegates.

The objective of CEAC was to provide a platform for engineers, scientists and industrial partners to discuss and disseminate recent research findings in wide aspects of theoretical, experimental, and practical civil engineering and architecture fields. Professionals, experts, and scholars from various engineering disciplines were invited to share experiences, innovations, achievements and knowledge. The conference brought together academics, researchers, and practitioners with the aim of bridging the gap between theory and practice in all civil engineering aspects. All papers in this proceeding were subject to peer-review by conference committee members and international reviewers. The papers were selected for the proceedings based on quality and relevance to the conference.

We express our gratitude to the Organising Committee for its valuable contribution to this successful edition of CEAC 2021. We also express our sincere gratitude to all people who contributed to the review process and did all their best to enhance the scientific merit and quality of the Proceedings. We are also grateful to the authors for their contributions, to the speakers and participants of the CEAC 2021. We wish to extend our warmest welcome to all interested participants to the next CEAC 2022. We also address thanks to all those involved in organising this edition of the conference, as well as to IOP Conference Series: Materials Science and Engineering.

Prof. C. W. Lim
City University of Hong Kong, Hong Kong

# CEAC Committees

## Conference Chairs

Prof. Tan Kiang Hwee, National University of Singapore, Singapore
Prof. C. W. Lim, City University of Hong Kong, Hong Kong

## Program Chairs

Prof. Kyoung Sun Moon, Yale University, USA
Prof. Ippei Maruyama, Nagoya University & The University of Tokyo, Japan

IOP Conf. Series: Materials Science and Engineering 1165 (2021) 011001 doi:10.1088/1757-899X/1165/1/011001

## Publicity Committee

Assoc. Prof Galina Gorbacheva, Bauman Moscow State Technical University, Russia

## Technical Committees

Prof. Naeem Ejaz, University of Engineering & Technology Taxila, Pakistan
Prof. Luigi Coppola,University of Bergamo, Italy
Prof. Osama Ahmed Mohamed， Abu Dhabi University, United Arab Emirates
Prof. Abeer Samy Yousef Mohamed, Effat University, Saudi Arabia- Tanta University Egypt
Assoc. Prof.Sarunya Promkotra,Khon Kaen University,Thailand
Dr. Trong-Phuoc Huynh, Can Tho University, Vietnam
Dr.Tamilsalvi Mari,Taylor's University Lakeside Campus,Malaysia
Dr. Koorosh Gharehbaghi, RMIT university, Australia
Dr. Fernando Vegas López-Manzanares, Universitat Politècnica de València, Spain
Dr. Ali Sarrafi Nik,Islamic Dr. Azad University,Iran
Dr. Francesca Scalisi, DEMETRA Ce.Ri.Med. (Euro-Mediterranean Documentation and Research Center), Italy
Dr. SITI NAZAHIYAH RAHMAT,Universiti Tun Hussein Onn Malaysia,Malaysia
Lecturer Attasit Sirivachiraporn,Srinakharinwirot University, Thailand
Koorosh Gharehbaghi, RMIT University, Australia
Omid Reza Baghchesaraei, Western Sydney University, Australia
HAGE CHEHADE Fadi, Lebanese University, Lebanon

IOP Conf. Series: Materials Science and Engineering 1165 (2021) 011002 doi:10.1088/1757-899X/1165/1/011002

# Peer review declaration

All papers published in this volume of *IOP Conference Series: Materials Science and Engineering* have been peer reviewed through processes administered by the Editors. Reviews were conducted by expert referees to the professional and scientific standards expected of a proceedings journal published by IOP Publishing.

- **Type of peer review:** Double-blind
- **Conference submission management system:** iConf Conference System
- **Number of submissions received: 20**
- **Number of submissions sent for review:** 20
- **Number of submissions accepted:** 11
- **Acceptance Rate (Number of Submissions Accepted / Number of Submissions Received X 100):** 55%
- **Average number of reviews per paper:** 2
- **Total number of reviewers involved:** 10
- **Any additional info on review process:**
  All received submissions were subjected to double-blind peer reviews, who are expert or have been experiencing in the related field for years. The papers are evaluated in terms of novelty, technical content, evaluation of experimental results, and presentation of the topics.

  The accepted papers must be revised according to the referees' comments and suggestions, before inclusion in the conference proceedings. Manuscripts submitted should be original & unpublished and written in accordance with the standard IOP paper template.
- **Contact person for queries:** Ms. Violet Xie (ceac@materials.ac.cn)

IOP Conf. Series: Materials Science and Engineering 1165 (2021) 012001 doi:10.1088/1757-899X/1165/1/012001

# Surface Observation of Laser-treated Carbon Steel and Application of Laser Ccleaning for Corroded Steel Member

**S Zhuang[1,*], S Kainuma[2], M Yang[2], M Haraguchi[3], T Asano[4]**

[1] Department of Urban and Environmental Engineering, Kyushu University, Fukuoka, Fukuoka 819-0395, Japan
[2] Department of Civil Engineering, Faculty of Engineering, Kyushu University, Fukuoka, Fukuoka 819-0395, Japan
[3] CoolLaser Division, TOYOKOH Inc, Hamamatsu, Shizuoka 434-0004, Japan
[4] Bridge and Structural Engineering Division, Technical Development and Environment Department, Technical Development Bureau, West Nippon Expressway Co., Ltd., Kita-ku, Osaka 530-0003, Japan

zhuang-ss@doc.kyushu-u.ac.jp

**Abstract.** A newly developed laser system is adopted in this study as a surface preparation method for corroded steel structures. To investigate the thermal effects of laser beam on steel material, carbon steel plates are treated by two different laser processing parameters. Then the surface condition, heat-affected zone, and iron oxide on laser-treated specimens are evaluated using optical microscope, SEM-EDX, and XRD. Moreover, a severely corroded steel member was used as laser irradiation target to verify the cleaning effects of laser beam. Results show that surface morphologies are varied according to laser processing parameters. Due to thermal effects, iron oxide was formed after laser irradiation, and higher laser power induced deeper heat-affected zone. The application of laser treatment for corroded steel member showing desirable surface cleanliness after laser irradiation. The laser surface treatment could be an alternative surface preparation method for steel structures in the construction field.

## 1. Introduction

The current surface preparation methods used in construction field are lacking ability to remove residual salts inside corrosion pits, or reach a complete cleaning state for severely corroded steel structures. Thereby, the underlying contaminates induce corrosion beneath coating, then reduce its durability [1]. To overcome these problems, laser treatment was introduced in this study as a promising surface preparation method for acquiring more desirable surface condition. Laser had been drawing attention in automotive and shipbuilding area, as a practical surface preparation method with extraordinary cleaning efficiency [2]. The contaminates on steel surface can be removed completely under appropriate laser conditions [3]. A high-power continuous wave (CW) laser was adopted in this study for steel surface preparation. Surface morphology and oxide composition were identified using scanning electron microscope-energy dispersive X-ray spectrometry (SEM-EDX) and X-ray diffraction (XRD). Moreover, a steel member suffered from severe corrosion problem was treated by laser cleaning as an application. The surface conditions were analyzed after laser irradiation, to evaluate the effectiveness of laser surface cleaning.

## 2. Experimental

### *2.1. Laser cleaning and surface observation*

Laser equipment adopted for surface treatment is a newly developed 3 kW CW fiber laser system. To enhance the treatment efficiency, the laser beam is rotated by a prism as a halo-like laser ring. Then project the laser ring on target surface and scan the whole area in given velocity, as shown in Figure 1. The total output energy during laser treatment divided by irradiated area was defined as energy density, increasing energy density represent higher heat effects. In surface characteristic analysis, the applied laser treatments including two different energy densities of 600 and 1282 J/cm$^2$, then labeled as Laser-C and Laser-D respectively. Carbon steel plate according to JIS G3106 SM490A with dimensions of 150×70×6 mm is used as specimen for surface characteristics. In consideration of identical surface condition before laser treatment, the steel plate was pretreated by abrasive blasting to remove mill scale on the surface. On the other hand, milling was carried on another group of abrasive blasted steel plate to remove the residual blasting material, also used as a comparison to laser surface treatment. SEM-EDX study was performed on specimen surface to identify the element distribution of Iron and Oxygen after laser irradiation. XRD measurements were carried out with generator settings of 40 kV and 30 mA, and the data was collected over a range of 20° to 90°.

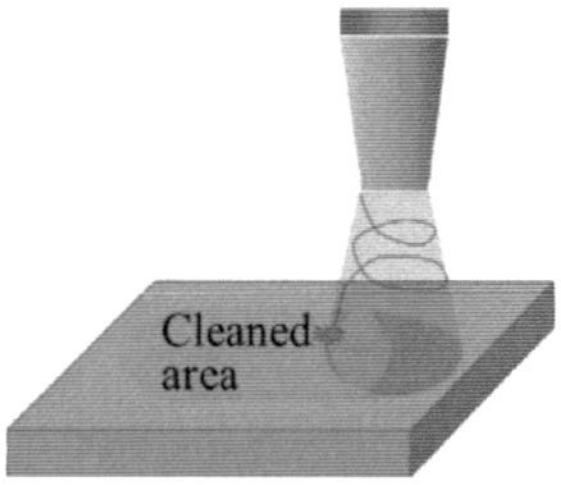


**Figure 1.** Schematic image of laser cleaning.

**Figure 2.** Severely corroded steel member.

### *2.2. Application of laser surface cleaning*

During the application test, the higher laser density condition (Laser-D, 1282 J/cm$^2$) was selected. A severely corroded steel member removed from rock shed was prepared for application of laser surface cleaning, as shown in Figure 2. The rock shed was exposed at coastal region of Kagoshima (Lat. 31°53′51″N, Long. 130°13′20″E) for 41 years, an 0.03 m$^2$ area was chosen as cleaning target. Before laser irradiation, rotary bristle tool was used as pretreatment to remove the outer rust layer in consideration of reducing laser energy wastage. The principle for laser irradiation is to remove any visible corrosion products on the surface in minimal treatment period. Surface roughness and salinity were tested to verify the effectiveness of laser surface cleaning.

## 3. Test results

### *3.1. Surface observation of laser-treated carbon steel*

Figure 3 shows the optic microscope of surface treated specimens. The laser irradiated steel plates tend to have rougher surface features than milling specimen due to laser ablation effects. Laser scanning paths were observed in Laser-D specimen, yet Laser-C presented no distinct scanning paths except for some craters. Laser-D possessed a relatively higher output power than Laser-C, thus the ablation would be enhanced, resulting in more clear appearance of laser scanning paths. Laser irradiation also has significant impact on transform the microstructure of metallic materials [4]. Especially for Laser-D condition, the higher energy density affects microstructure of near surface steel layer, and causing deeper heat-affected zone (HAZ) on the surface [5], as shown in Figure 4. However, the HAZ of Laser-C specimens is unrecognizable under the same magnification.

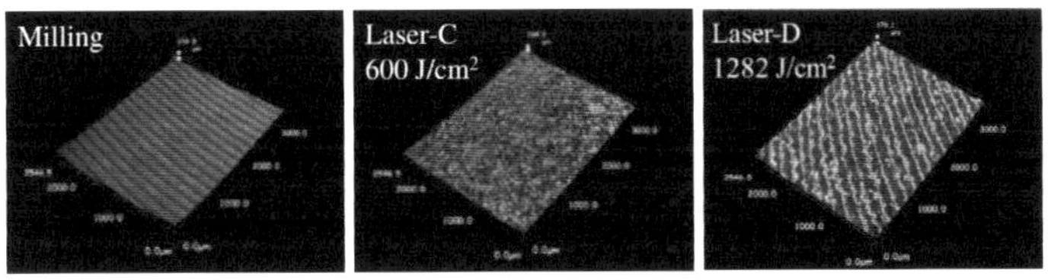


**Figure 3.** Surface morphology of treated specimens.

**Figure 4.** HAZ evaluation on the cross-section.

From SEM-EDX results shown in Figure 5, oxygen is observed on the steel surface after laser treatment. In which the oxygen mainly existing inside laser ablation craters of Laser-C, and the edge of scanning paths of Laser-D. Semi-quantitative analysis of EDX reveals that oxygen on the laser-treated surface achieved 9.72 mass% and 4.87 mass% for Laser-C and Laser-D specimens respectively, while milling steel surface possess 1.03 mass% of oxygen. The principle of laser ablation is based on melting and vaporization. When the laser beam hits steel surface, the material absorbed laser energy and start to melt or vaporize [6]. Larger energy density causes higher temperature on the heated material, therefore the ablation is enhanced, resulting in larger volumes of removed material from steel surface as mentioned above. As for Laser-C conditions, the laser processing mechanism relatively dominated by melting effects rather than ablation due to lower energy density. However, the temperature might be high enough to oxidize the steel surface before cooling, thus higher proportion of oxygen will be found on the surface. On the other hand, the energy density of Laser-D is large enough to produce a relatively ablation dominated processing mechanism [7]. The heated materials in higher temperature were ablated, and took away most of the energy by themselves. Therefore the remaining energy in the base metal could be relatively small, resulting clear laser paths and less oxygen on the surface. XRD results shown in Figure 6 revealed that wustite formed on the steel surface after laser irradiation, in which Laser-D specimen presented lower oxide intensity than Laser-C due to ablation dominated processing.

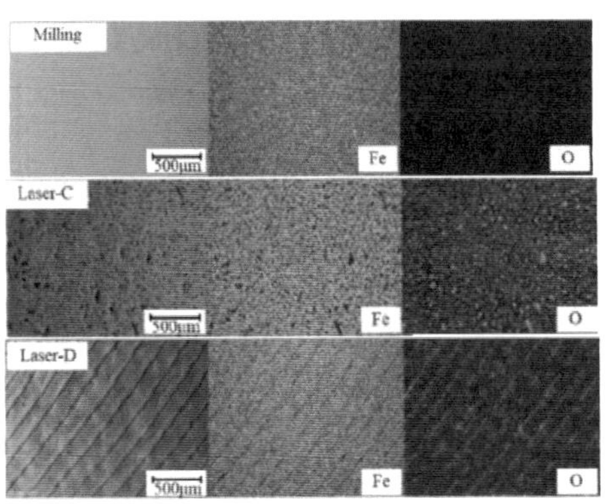


**Figure 5.** SEM-EDX results of specimens.

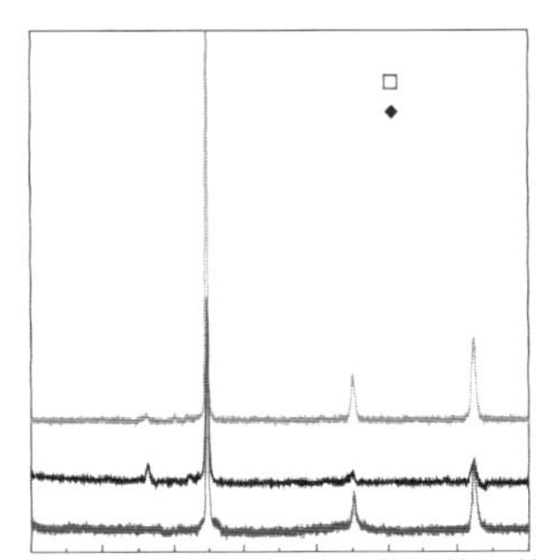


**Figure 6.** XRD diffraction patterns of specimens.

*3.2. Application of laser surface cleaning*

Figure 7 shows the surface condition pretreated by rotary bristle tool. Outer corrosion layers were removed from the steel surface, an irregularity appearance is revealed due to existing of large corrosion pits. Parts of the base metal are exposed and showing metal lusters. Higher magnification photos of the treated area confirmed corrosion products remaining on the steel surface. From the surface salinity result reported in Table 1, the residual salinity after pretreatment exceeded the limit of 50 mg/m$^2$ demanded by standard [8]. From surface condition after laser cleaning shown in Figure 7, corrosion products had been removed from the surface including some of the rust remained inside corrosion pits. Higher magnification photo presents a clean surface status without visible corrosion products. Clear intersected laser paths on the irradiated surface were observed, while some areas showing flat appearance without laser paths. Surface salinity after laser irradiation tends to meet the requirement of standard, reached a significantly lower amount of 1.30 mg/m$^2$. Besides, the surface roughness ($R_{ZJIS}$) after laser cleaning is 37.1 μm, satisfied the standard requirement of less than 80 μm. However, iron oxide on the laser-treated steel surface may affect the corrosion properties and coating quality, and future study is demand to evaluate the coating and corrosion resistance of laser-treated carbon steel.

**Figure 7.** Surface condition after treatment.

**Table 1.** Effectiveness of laser surface cleaning.

| Treatment | Surface salinity (mg/m$^2$) | Surface roughness $R_{ZJIS}$ (μm) |
|---|---|---|
| Pretreatment | 67.2 | 32.1 |
| Laser | 1.30 | 37.1 |

## 4. Conclusions

This study proposed a new surface preparation method using high-power CW laser. The effects of laser irradiation on steel surface and the practicability of laser cleaning were evaluated through various test methods. The results indicate:

1) The surface morphology of laser-treated steel depends on the laser processing parameter. In general, higher laser energy density results in more obvious laser irradiation features. Iron oxide formed on the laser-treated steel surface, the corrosion properties and coating quality on laser-treated carbon steel could be affected by the iron oxide.
2) Application test manifests the effectiveness of laser cleaning for corroded steel member. In which an extremely low surface salinity and desirable surface roughness were achieved.

## Acknowledgments

This research work was kindly supported by JSPS KAKENHI under Grant Numbers JP19K15074 and JP19H02227. The financial support from the Kyushu Regional Management Service Association is acknowledged. The support provided by the China Scholarship Council (File No.201809350010) is acknowledged.

## 5. Reference

[1] Kainuma S, Yang M, Ishihara S, Kaneko A and Yamauchi T 2019 Corrosion protection of steel members using an Al-Zn base sacrificial anode and fiber sheet in an atmospheric environment *Constr. Build. Mater.*

[2] Madhukar Y K, Mullick S, Shukla D K, Kumar S and Nath A K 2013 Effect of laser operating mode in paint removal with a fiber laser *Appl. Surf. Sci.* **264** 892–901

[3] Chen G X, Kwee T J, Tan K P, Choo Y S and Hong M H 2012 High-Power Fibre Laser Cleaning for Green Shipbuilding **7** 249–54

[4] Min J, Wan H, Carlson B E, Lin J and Sun C 2020 Application of laser ablation in adhesive bonding of metallic materials: A review *Opt. Laser Technol.*
[5] Liu S and Shin Y C 2019 Additive manufacturing of Ti6Al4V alloy: A review *Mater. Des.*
[6] Zhuang S, Kainuma S, Yang M, Haraguchi M and Asano T 2021 Investigation on the peak temperature and surface defects on the carbon steel treated by rotating CW laser *Opt. Laser Technol.* **135** 106727
[7] Schmidt M, Merklein M, Bourell D, Dimitrov D, Hausotte T, Wegener K, Overmeyer L, Vollertsen F and Levy G N 2017 Laser based additive manufacturing in industry and academia *CIRP Ann.*
[8] Anon 2014 *Steel bridge anticorrosion manual* (Japan Road Association)

# Shear Strength of SFRC Piles without Shear Reinforcement using the Centrifugal Forming Technique

**R A Raju[1], M Akiyama[1]*, S Lim[1], T Kakegawa[1], Y Hosono[1], and Y Uno[2]**

[1] Department of Civil and Environmental Engineering, Waseda University, Tokyo 169-8555, Japan
[2] Sato Kogyo Co., Ltd., Kanagawa 243-0123, Japan

akiyama617@waseda.jp

**Abstract.** This paper proposed a fabrication method of steel fiber reinforced concrete (SFRC) piles without shear reinforcement using the centrifugal forming technique to reorient the fibers perpendicular to the pile's longitudinal axis for resisting the shear stresses. Effects of rotation patterns and concrete mixes on the fiber dispersion were investigated experimentally to identify the optimum procedure providing both well-oriented and well-distributed steel fibers. Using the X-ray radiography to visualize steel fibers inside the piles, it is confirmed that the centrifugal forming technique can realign the fibers in favor of the shear stresses. The shear test results show the effectiveness of the steel fibers in improving the shear strength of the SFRC piles due to good fiber dispersion. Moreover, the SFRC piles without shear reinforcement exhibited more ductile behavior than the RC piles with shear reinforcements. Steel fibers can be used to replace the shear reinforcements and improve the demanding productivity in the pile construction industry by saving time and labor.

## 1. Introduction

The centrifugal forming hollow circular reinforced concrete (RC) piles have a greater flexural capacity yet lower shear strength compared to the traditional RC piles [1]. Although the higher shear reinforcement ratio can increase the shear strength, excessive large crack-widths and concrete spalling are major disadvantages for the RC piles [2]. The benefit of adding steel fibers to concrete can be achieved in terms of enhanced post-cracking performance due to the crack-bridging fibers [3]. However, previous studies (e.g., [4]) demonstrated that the fiber distribution and orientation in steel fiber reinforced concrete (SFRC) are random and difficult to control. This impairs the effectiveness of using the fibers to enhance the structural performance of SFRC piles [5].

The centrifugal forming technique can be a solution for improving fiber orientation and enhancing the shear strength of the SFRC piles. The main concept is that the pile rotation by the centrifugal forming method would generate friction forces between the pile forms and SFRC, which could influence the fibers to change their alignment perpendicular to the pile's longitudinal axis enhancing the piles' shear resistance. If this task can be achieved, shear reinforcements can be replaced with steel fibers in centrifugally formed piles, and the pile construction method will be less laborious.

This paper presents a novel fabrication method for the centrifugal forming SFRC piles without shear reinforcements and investigates their shear strength. An experimental procedure was carried out to investigate the effect of rotation patterns and concrete types on the fiber distribution and orientation in the SFRC piles using X-ray radiography. After obtaining the optimum casting procedure, the SFRC piles without shear reinforcement were fabricated for loading test, and their shear strength was compared with those of the RC piles with shear reinforcements.

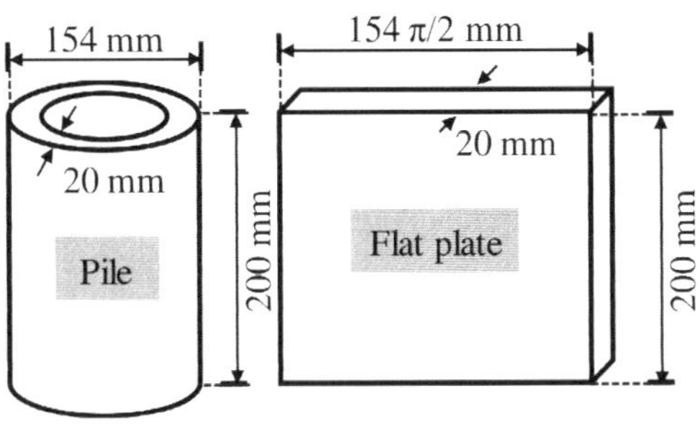


**Figure 1.** Test specimens.

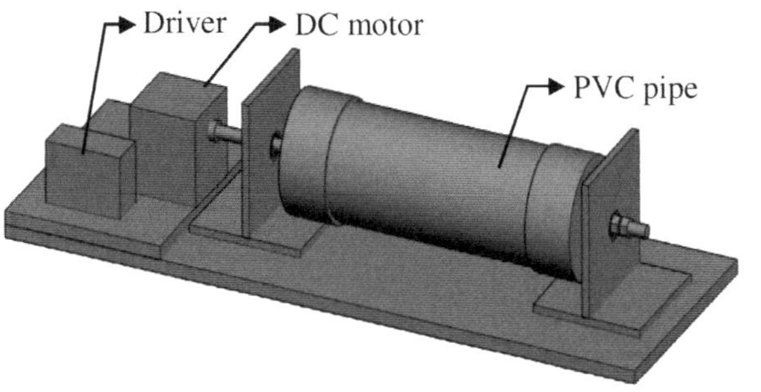


**Figure 2.** Centrifugal forming method for piles.

**Table 1.** Rotation patterns for the centrifugal forming.

| rpm (revolutions/minute) | 83 | 183 | 323 | 481 | 590 |
|---|---|---|---|---|---|
| R1 (minutes) | 5 | 3 | 3 | 3 | 5 |
| R2 (minutes) | 10 | 3 | 3 | 2 | 2 |

**Table 2.** Concrete mix proportions.

| Mix type | Water ($kg/m^3$) | Cement ($kg/m^3$) | Limestone ($kg/m^3$) | FA ($kg/m^3$) | CA ($kg/m^3$) | SP ($kg/m^3$) | AE ($kg/m^3$) | Steel fiber ($kg/m^3$) | Slump (mm) |
|---|---|---|---|---|---|---|---|---|---|
| SCC | 175 | 318 | 224 | 853 | 708 | 7.86 | 1.08 | 0.50 | 525 |
| NC1 | 175 | 350 | 0 | 950 | 789 | 3.50 | 0.70 | 0.50 | 42 |
| NC2 | 175 | 350 | 0 | 1036 | 701 | 5.25 | 0.70 | 0.50 | 100 |

Note: FA = fine aggregate, CA = coarse aggregate, SP = superplasticizer, and AE = air-entrained admixture.

## 2. Development of fabrication process of centrifugally formed SFRC piles

### *2.1. Test specimens and casting method*

Figure 1 shows the test specimens of the circular hollow pile and flat plate. The effects of the rotation patterns and concrete mixes on the fiber distribution and orientation are investigated herein. Table 1 shows two different rotation patterns used to fabricate the piles, denoted as "R1" and "R2". The plates were made without centrifugal forming method, denoted as "R0". Table 2 presents three different concrete mix proportions used to fabricate the specimens. The SCFRC mix is denoted as "SCC", and two SFRC mixes with different fluidity are denoted as "NC1" and "NC2". Figure 2 presents the equipment setup of the centrifugal forming method to fabricate the piles using a polyvinyl chloride (PVC) pipe as formwork and a DC motor and driver to control the rotation speed. The coarse aggregate with a maximum size of 10 mm was used in the concrete mix. A steel fiber (RC-65/35-BN) with single hook-ends, a length of 35 mm, and a diameter of 0.55 mm was used.

### *2.2. X-ray photography and fiber measurement method*

Figure 3 shows the setup for the X-ray photography for each specimen. The X-ray images were captured from the piles' outer surface from every 45° viewing angle and the plates' lateral side. Then, the fibers were measured and counted from the X-ray images to obtain each fibers position and angle (α), as shown in figure 4. To investigate the steel fiber distribution and orientation in the specimens, each X-ray image was divided into 29.5 x 45 mm meshes, and the fiber amount in each mesh was calculated. Then, the coefficient of variation ($CoV_f$) was determined by dividing the average by the standard deviation associated with the number of steel fibers among the meshes. $CoV_f$ is used as a fiber distribution index, where the lower $CoV_f$ values mean uniform fiber distribution. Figures 5 and 6 show the variation of the fiber amount oriented in different angles ranging from 0° to 180° for the test specimens, where the 90° angle represents perpendicular to the piles' longitudinal axis.

IOP Conf. Series: Materials Science and Engineering 1165 (2021) 012002 doi:10.1088/1757-899X/1165/1/012002

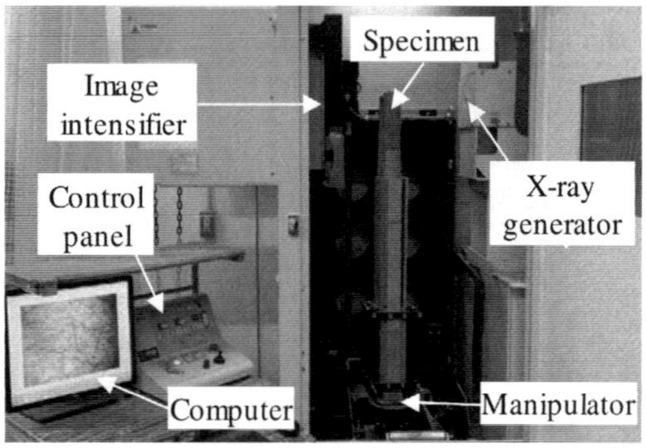


**Figure 3.** Setup for the X-ray photography.

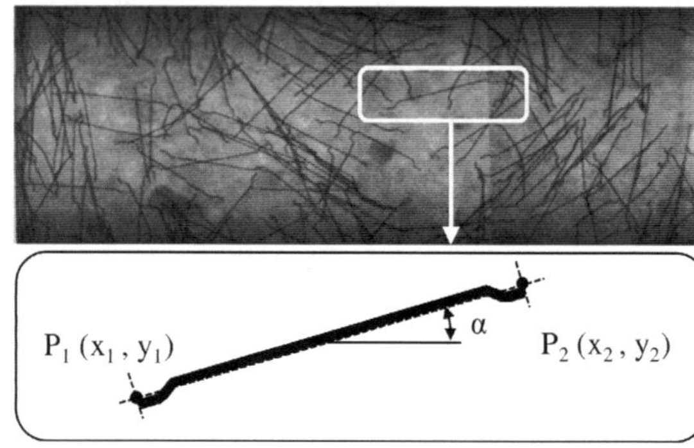


**Figure 4.** Fiber counting from X-ray images.

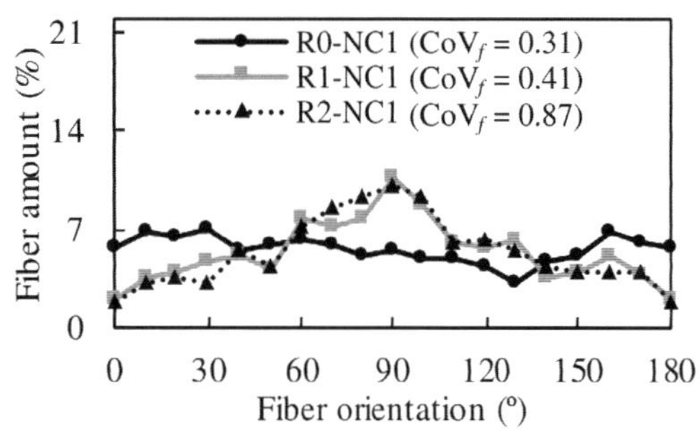


**Figure 5.** Effect of rotation pattern.

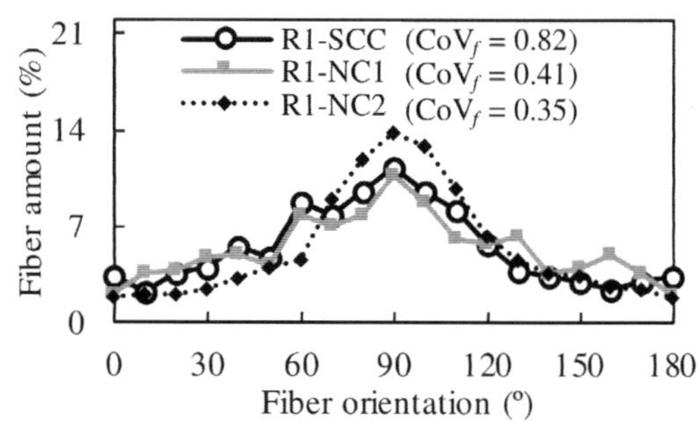


**Figure 6.** Effect of concrete type.

### *2.3. Effect of rotation pattern and concrete type on the fiber distribution and orientation*

Figure 5 presents the effect of different rotation patterns of centrifugal forming on the fiber orientation and distribution in specimens R0-NC1, R1-NC1, and R2-NC1. Figure 5 shows that a significant number of fibers are oriented close to 90° angle in the piles fabricated with the centrifugal forming method (i.e., R1-NC1 and R2-NC1) regardless of the rotation patterns compared to the randomly oriented fibers in the plate (i.e., R0-NC1). In contrast, the fiber distribution was less uniform in the piles with different intensities, depending on the rotation patterns. It should be noted that the rotation duration at the beginning and end of the centrifugal forming method contributes to the concrete material distribution and compaction around the pile's circumference, respectively [6]. Figure 5 confirms that rotation pattern R1 (i.e., $CoV_f$ of R1-NC1 = 0.41) produced better fiber distribution than rotation pattern R2 (i.e., $CoV_f$ of R2-NC1 = 0.87).

Figure 6 presents the fiber orientation and distribution in the SCFRC (R1-SCC) and SFRC (R1-NC1 and R1-NC2) piles. It can be seen that the SFRC piles produced better fiber orientation and distribution ($CoV_f$) than those in the SCFRC pile. The reason might be attributed to the shorter hardening time of the SCFRC that made it difficult for the fibers to reorient by the centrifugal forming method. Moreover, the higher fluidity of SCFRC dispersed the fibers during the pile rotation. As for the SFRC piles, similar fiber distribution (i.e., $CoV_f$ = 0.35 and 0.41) is obtained, and R1-NC2 produced better fiber orientation than R1-NC1.

Considering the outcomes shown in Figures 5 and 6, the NC2 mix and the rotation pattern R1 were used to manufacture the centrifugal forming piles with longitudinal rebars for the shear test.

## 3. Shear test of the centrifugal forming piles

### *3.1. Details of pile specimens and loading test*

Table 3 lists the details of the centrifugal forming piles designed for this study. Figure 7 shows the loading and supporting points, the rebar arrangements, and the strain gauge positions in the piles. The SFRC piles had the steel fiber content ($v_f$) of 0.50% and 1.00%, and two piles were prepared for each series to consider the variation in the fiber distribution and orientation of each pile.

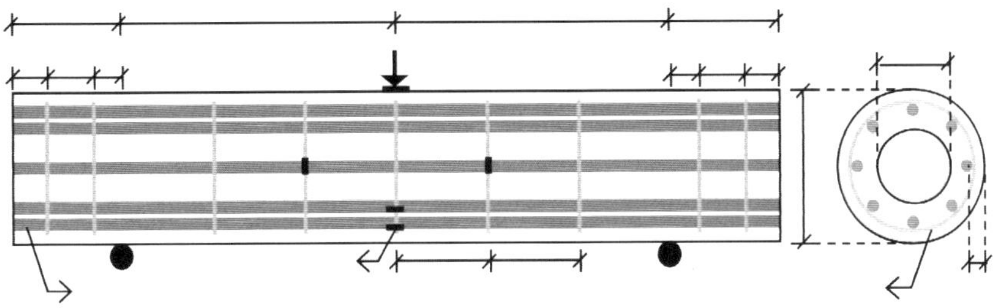

**Figure 7.** Details and loading method of centrifugal forming piles (units in mm).

**Table 3.** Details of the centrifugal forming piles and the shear test results.

| Pile notation | Steel fiber content, $v_f$ (%) | Shear reinforcement spacing, $s$ (mm) | Shear reinforcement ratio, $\rho_w$ (%) | Compressive strength, $f'_C$ (N/mm$^2$) | Tensile strength, $f_t$ (N/mm$^2$) | Peak load, $P_U$ (kN) | Failure mode |
|---|---|---|---|---|---|---|---|
| PF0[a] | 0 | 0 | 0 | 47.7 | 3.83 | 76 | Shear |
| PS1[b] | 0 | 142 | 0.22 | | | 86 | |
| PS2[b] | 0 | 95 | 0.30 | | | 99 | |
| PF50-1[c] | 0.50 | 0 | 0 | 47.5 | 4.48 | 107 | |
| PF50-2[c] | 0.50 | 0 | 0 | | | 90 | |
| PF100-1[c] | 1.00 | 0 | 0 | 47.6 | 5.47 | 129 | Bending |
| PF100-2[c] | 1.00 | 0 | 0 | | | 118 | |

[a] Control pile. [b] RC pile. [c] SFRC pile.

*3.2. Shear strength of control and RC piles*

Table 3 presents the shear test results for the piles. Figure 8a shows the load-deflection (P-δ) responses of control and RC piles. Without the shear reinforcement, the control pile PF0 attained the peak load ($P_U$) of 76 kN. Whereas the RC pile PS2 with the larger shear reinforcement ratio (i.e., $\rho_w$ = 0.30%) exhibits a higher $P_U$ of 99 kN compared to the RC pile PS1 (i.e., $\rho_w$ = 0.22%) with $P_U$ of 86 kN. For piles PF0 and PS1, brittle behavior was observed with a sudden load drop after attaining the peak load, while less brittle behavior was observed for PS2. Figure 8c shows the cracking pattern at the failure of pile PS2, where large diagonal cracks (thick lines) with severe concrete damage were observed. Finally, the shear failure mode was observed for all control and RC piles.

*3.3. Shear strength of SFRC piles*

Figure 8b depicts the P-δ responses of SFRC piles (i.e., PF50 and PF100 series) without shear reinforcements. The addition of $v_f$ = 0.50% (i.e., PF50 series) caused a noticeable increase in the $P_U$ of 107 and 90 kN compared to the control pile (i.e., $P_U$ = 76 kN). The difference of $P_U$ between the PF50 series might be attributed to the variation of fiber dispersion. Both piles PF50-1 and PF50-2 exhibit shear failure mode since none of the tensile rebars were yielded. However, they did not fail in a brittle manner, and no rapid load reduction was observed.

Moreover, the PF100 pile series containing $v_f$ = 1.00% displays significantly higher loading capacities of 129 and 118 kN with ductile behavior. The yielding of the tensile rebars was observed for pile PF100-1 just before attaining its peak load, and the flexure failure mode was confirmed. For pile PF100-2, all of the strain gauges of the tensile rebars were damaged during casting. However, the failure mode of pile PF100-2 was determined as flexure failure due to its similar P-δ response to that of pile PF100-1, and its peak load (i.e., $P_U$ = 118 kN) was close to the bending capacity of 114 kN estimated by ACI 318-11 [7]. Therefore, it is apparent that the failure mode transition from shear to flexure failure was observed for the SFRC piles with 1.00% of steel fibers.

Regarding the crack propagation, the SFRC piles exhibited better crack behavior due to the crack-bridging fibers compared to the RC piles. For example, Figures 8c and 8d demonstrate that the SFRC pile PF100-1 has fewer and smaller cracks than the RC pile PS2 at the failure stage.

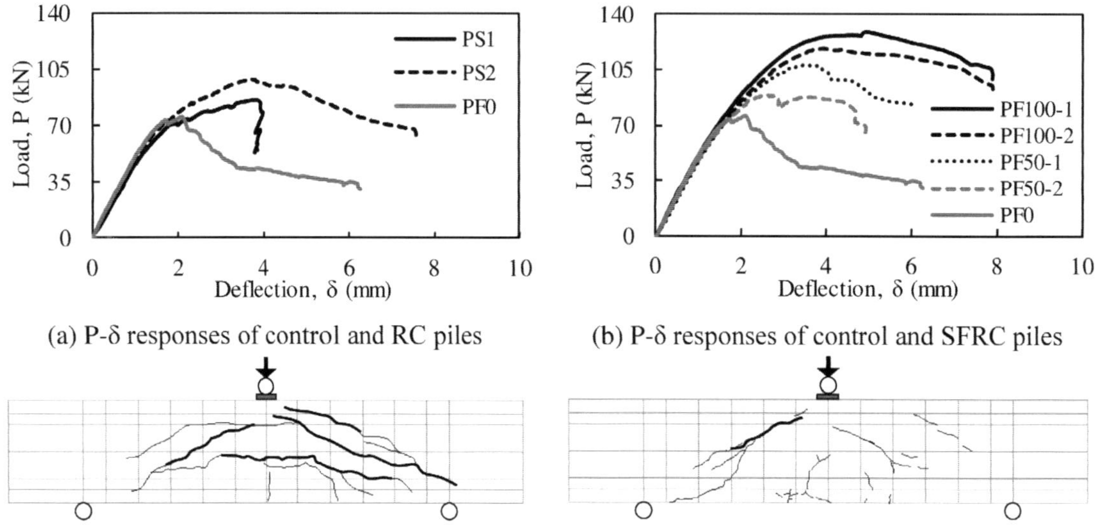


(a) P-δ responses of control and RC piles (b) P-δ responses of control and SFRC piles

c) Crack pattern of RC pile PS2 (d) Crack pattern of SFRC pile PF100-1

**Figure 8.** P-δ responses and crack patterns at the ultimate condition of centrifugal forming piles.

## 4. Conclusion

In this study, the fabrication method of SFRC piles using the centrifugal forming technique is proposed to improve the fiber distribution and orientation, as visualized by X-ray photography. The shear test results confirmed that the shear strength of the centrifugal forming SFRC piles is increased significantly with high steel fiber contents. Moreover, the SFRC piles exhibit more stable crack propagation with smaller crack widths than the RC piles. Therefore, it is concluded that steel fibers can replace the shear reinforcements in the centrifugal forming SFRC piles.

## Acknowledgments

This work was supported by JSPS KAKENHI grant number 19H00813.

## 5. References

[1]Jensen U G and Hoang L C 2010 Shear strength of reinforced concrete piers and piles with hollow circular cross section *Struct. Eng. Int.* **20** 260–7

[2]Akiyama M, Abe S, Aoki N and Suzuki M 2012 Flexural test of precast high-strength reinforced concrete pile prestressed with unbonded bars arranged at the center of the cross section *Eng. Struct.* **34** 259–70

[3]Narayanan R and Darwish I Y S 1987 Use of steel fibers as shear reinforcement *ACI Struct. J.* **84** 216–27

[4]Raju R A, Lim S, Akiyama M and Kageyama T 2020 Effects of concrete flow on the distribution and orientation of fibers and flexural behavior of steel fiber-reinforced self-compacting concrete beams *Constr. Build. Mater.* **262** 119963

[5]Lim S, Raju R A, Matsuda M, Okamoto T and Akiyama M 2018 Structural behavior prediction of SFRC beams by a novel integrated approach of X-ray imaging and finite element method *Constr. Build. Mater.* **170** 347–65

[6]Aya K 1960 On compacting concrete of reinforced concrete piles by centrifugal method (Japanese) *Trans. Japan Soc. Civ. Eng.* **71** 1–16

[7]ACI Committee 318 2011 Building code requirements for structural concrete (ACI 318-11) and commentary *Am. Concr. Inst.* 509

# Fundamental Study on Weather Resistance of Overlapping Layers between Al-5Mg Alloy Thermal Spray Coating and Heavy-duty Paint Coating

**H Yang[1,*], S Kainuma[2], M Yang[2], K Muto[3], T Asano[4]**

[1] Department of Urban and Environmental Engineering, Kyushu University, Fukuoka, Fukuoka 819-0395, Japan
[2] Department of Civil Engineering, Faculty of Engineering, Kyushu University, Fukuoka, Fukuoka 819-0395, Japan
[3] Technical Development Division, Fujigiken Co., Ltd, Osaka, Japan
[4] Bridge and Structural Engineering Division, West Nippon Expressway Co., Ltd., Osaka, Japan

yang-hx@doc.kyushu-u.ac.jp

**Abstract.** In this study, to clarify the weather resistance of Al-5Mg alloy thermal spray coating film, the effect of thermal spraying fabrication was considered. The accelerated weathering test with a hydrogen peroxide solution was performed on the overlapping layer specimens, which compared two types of fluororesin as the top coating of heavy-duty paint coating. The test results clarified the effect from thermal spraying fabrication had almost no effect on the weather resistance of the heavy-duty paint coating when the spraying process was applied to the top coating.

## 1.Introduction

Recently in Japan, Al-5Mg alloy thermal spraying coatings (AMT) have been used as an anti-corrosion coating applied for steel bridges, to improve the corrosion resistance in highly corrosive chloride-containing environments[1][2][3]. And it was partly covered by heavy-duty paint coating when it is hard to apply the thermal spray coating on the bridge components. These thermal spray duplex coating systems are supposed to provide a long lifetime, and also will be both cost-effective and environment friendly. However, in the previous study[1][2], these duplex coating systems showed low durability than thermal spray coating and severe film blister has occurred when defects were introduced into specimens.

Figure 1(a) shows the current method in which the heavy-duty paint coating is partially applied to the thermal spray coating after thermal spray coating was conducted. Figure 1(b) shows the contrast method with Figure 1(a), that applying the heavy-duty paint coating on the thermal spray coating. By applying this method, there is no overlapping layer with the heavy-duty paint coating on the thermal spray coating, so that the blister of the film according to the sacrificial anticorrosion effect is unlikely to occur. However, the effect of thermal spraying on the weather resistance of the coating is unknown. Therefore, to investigate the effect of thermal spraying on the weather resistance of this coating film for fluororesin paints, we conducted a hydrogen peroxide spray type accelerated weather resistance test (HP-Xe test)[4] to accelerate the deterioration of the paint coating. Furthermore, to quantitatively evaluate the film deterioration of the overlapped part of thermal spray coating and the paint coating

IOP Conf. Series: Materials Science and Engineering 1165 (2021) 012003 doi:10.1088/1757-899X/1165/1/012003

(overlapping region), and examine the weather resistance of the overlapping region, elemental analysis by SEM-EDX and gloss degree measurement were performed.

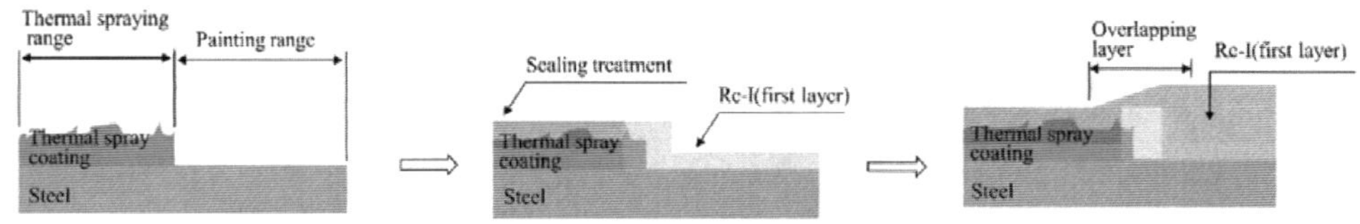


(a) Current method.

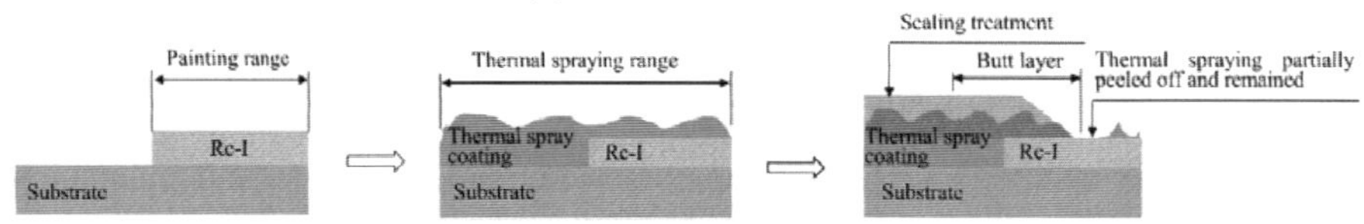


(b) Proposed method.

**Figure 1.** Countermeasure for early deterioration in the overlapping layer between thermal spray coating and heavy-duty coating.

## 2.Experimental procedure

*2.1Test Specimen*

The aluminum alloy plate with dimensions of 150×70×2 mm was used for the test, following Japanese Industrial Standards (JIS) H 4000 A1050. Figure 2 shows the shape and dimensions of the test piece and conditions of the coating film. For the paint, we used fluororesin, which is generally used for the topcoat of heavy-duty paint coating, and $TiO_2$, which has two types of coating treatment with different photocatalytic action. F1 is a highly weather-resistant fluororesin containing titanium oxide coated with Si treatment and Al chelate. On the other hand, F2 is a standard weather-resistant fluororesin containing titanium oxide coated with Si treatment. The test piece simulates the overlapping region between the coating film and the sprayed film shown in Figure 1 (b).

In the accelerated deterioration test of the coating film, which will be described later, a total of four types of specimens are performed: specimens that have been thermally sprayed onto the surface of the F1 and F2 paint film (F1 and F2), and F1 and F2 paint coating specimens. With these test specimens, the weather resistance of the coating film with and without thermal spraying was examined. Since the adhesion between the thermal spray coating and the coating film was low for both the F1 and F2 specimens, the surface layer of the thermal spray coating was partially peeled off immediately after the thermal spraying. Plasma arc spraying (air pressure: 0.40 N/mm$^2$, plasma gas pressure: 0.35 N/mm$^2$, current: 60A) was used for Al-5Mg thermal spraying. The temperature of the sprayed particles is about 200°C. The Al-5Mg alloy wire is A5056W (JIS H 4040) with a diameter of 1.6 mm. The target value for the thickness of the Al-5Mg thermal spray coating of each test piece was 200 μm.

*2.2HP-Xe test*

In JIS K 5659 regarding weather-resistant paints for steel structures, the weather resistance focusing on resins are evaluated based on the gloss retention rate of the coating film after 2,000 hrs of Xenon irradiation. However, since the main components of the paint are resin and pigment, it is necessary to consider the weather resistance of the pigment to properly evaluate the weather resistance of the coating film. In this study, the deterioration acceleration test of the fluororesin coating film by the photocatalytic action of the titanium oxide pigment in the coating film (HP-Xe test) was performed under the test conditions shown in Table 1[4][5][6]. For this test condition, a test condition in which a cycle of spray cleaning with pure water is repeated by combining spraying of hydrogen peroxide solution and irradiation with ultraviolet rays in an accelerated deterioration test tank equipped with a xenon arc lamp light source is used.

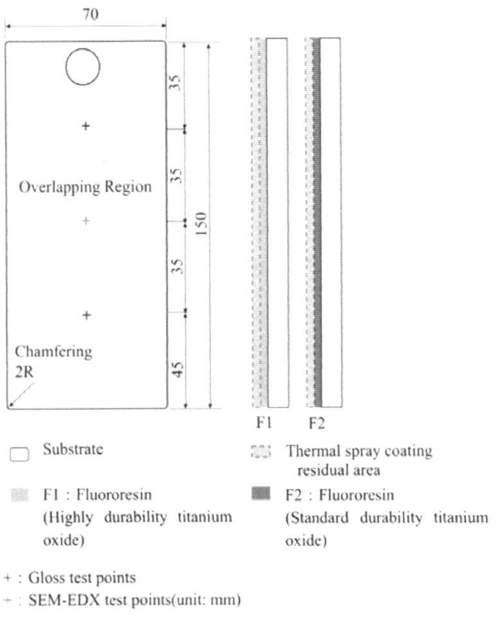


**Figure 2.** Dimensions of specimens and coating film conditions.

**Table 1.** Test conditions of HP-Xe test.

| Test conditions | Setting conditions |
|---|---|
| Lamp-house | Xenon arc lamp |
| Intensity of illumination | $80W/m^2$ |
| Lamp filter | Daylight |
| Temperature/humidity | 35°C / 70% |

*2.3Deterioration detection method*

For F1 and F2 specimens before and after the test, the surface of the specimens was observed by SEM at an acceleration voltage of 15 kV to identify the constituent elements of deposits on the film surface and evaluate their distribution. The resolution and acceleration voltage were respectively set to 3 nm and 30 kV. To confirm the elements on the surface of the overlapping region, SEM-EDX was performed to map Al, F, and Ti on the surface of the coating film.

The gloss of the test piece before and after the test was measured using a gloss meter following JIS K 5600-4-7. The measurement angle, measurement area, and measurement time were 60°, 9×15mm, and 0.5sec/angle, respectively. The measurement position of gloss was the position indicated by the + mark in Figure 2. The gloss was measured 11 times for each position, and the average value was defined as the gloss of each specimen.

**3.Test result**

Figure 3 shows the changes over time in the surface conditions of the F1 and F2 specimens after the HP-Xe test 200 hrs. In all the specimens, the thermal spray coating was partially peeled off from the coating film interface immediately after thermal spraying. After 50 hours, the hue of the sprayed coating film remaining on the coating film changed to gray. It has been reported that aluminum alloys may discolor in water. Also, the adhesion status of the sprayed metal of the F1 is the same as the initial state after 200 hours, and it adheres to the coating film. On the other hand, in the F2 specimen, the sprayed metal was peeled off immediately after the start of the test and disappeared almost entirely after 200 hours. It is considered that this is because the fluororesin around the titanium oxide pigment in the F2 specimen was decomposed and deteriorated by the photocatalytic action[5], and the adhesion between the thermal spray coating and the coating film was reduced. The element distribution and Al content on the surface of the film are shown in Figure 4 and Figure 5. During the test, the thermal spray coating deteriorated and peeled off, resulting in a decrease in the Al content and Ti, which is the pigment component of the paint. It is showed that this tendency of F2 is more prominent compared to

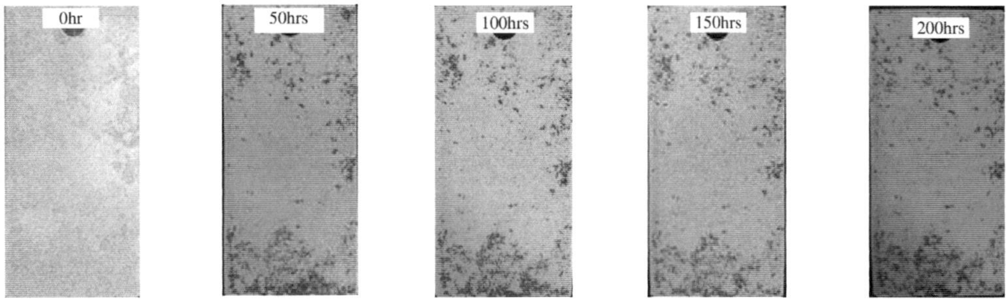


(a) F1

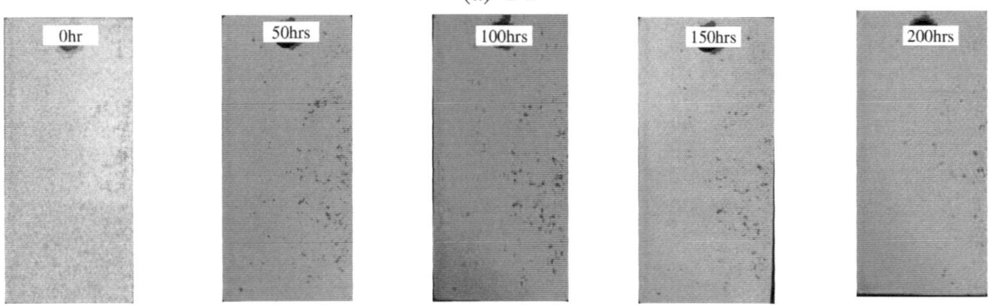


(b) F2

**Figure 3.** Photos of specimens over time.

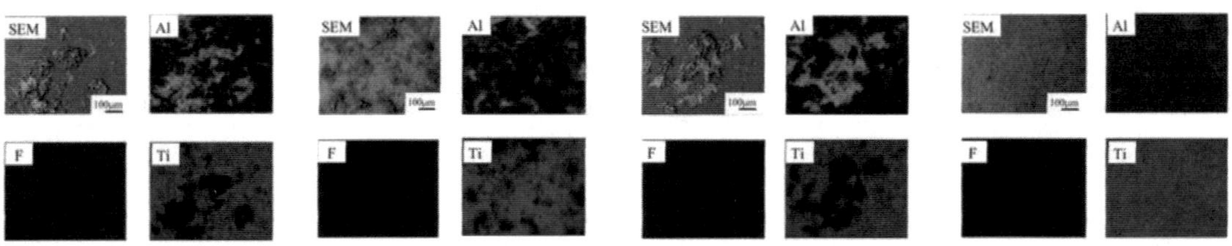


(a) F1(before tests) (b) F1(after tests) (c) F2(before tests) (d) F2(after tests)

**Figure 4.** Test results of SEM-EDX.

F1. As mentioned above, the deterioration of the fluororesin around the titanium oxide pigment of F2 was significantly worse than that of F1. Because of that, the adhesion between the thermal spray coating and the coating film was reduced. As a result, it is considered that the thermal spray coating was peeled off and almost disappeared entirely so that the coating film was completely exposed.

Figure 7 shows changes in the gloss retention rate X of the overlap specimen and paint coating specimen over time. Regardless of titanium oxide, the X of overlap specimen after 200 hrs is larger than that of paint coating specimen, and it is about 90% in the F1 specimen. Besides, since the X of the F1 specimen decreased except for about 20 hrs after the test started, it is inferred that the coating film is sound. Also, as the test time increased, the thermal spray coating on the coating film of the F2 specimen peeled off, and X decreased significantly, but it became almost constant after 50 hrs. This tendency is the same for F2 paint coating specimens. Since the F1 specimen has better weather resistance than the F2 test specimen. The gloss of the F1 specimen gradually decreases, and the gradient is small. The decrease in the gloss of overlap specimen is smaller for the F2 specimen than for the F2 paint coating specimen. However, the F2 specimen has lower weather resistance due to the photocatalytic action of titanium oxide than F1. Therefore, the effect of ultraviolet rays on X is small, and the gloss of the F2 paint coating specimen and the F2 specimen tends to decrease in the same way. Therefore, it can be said that thermal spraying has almost no effect on the weather resistance of the topcoat.

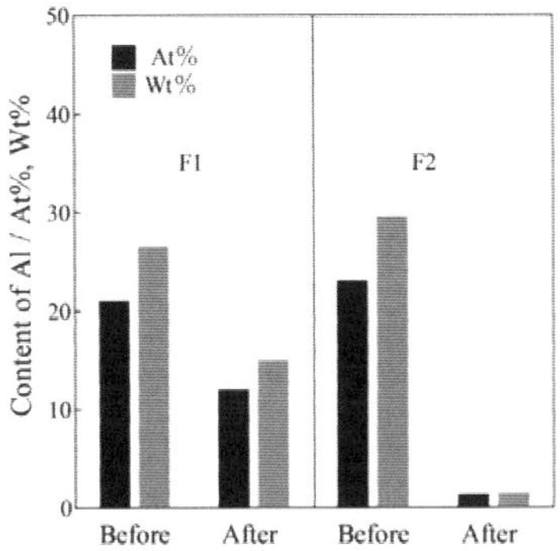


**Figure 5.** Content of Al on the surface of the specimen.

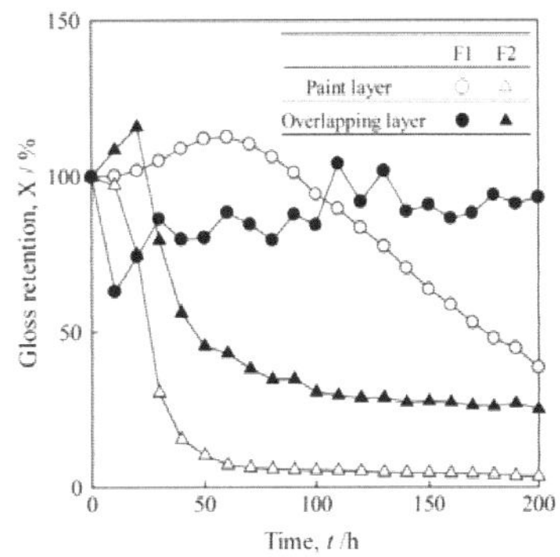


**Figure 6.** Changes in gloss retention over time.

## 4.Conclusions

In this study, to investigate the effect of thermal spraying on the weather resistance of the paint coating film, and improve the durability of the boundary between the Al-5Mg thermal spray coating and the heavy-duty paint coating, the HP-Xe test was conducted on the coating film. Besides, to examine the weather resistance of the overlap region between the thermal spray film and the paint coating film, SEM-EDX and gloss measurements were performed. The main results obtained in this study are shown below.

1) The weather resistance of the paint coating when thermally sprayed on a fluororesin paint coating with Si treatment and Al chelate treatment is higher than that of the fluororesin paint coating only with Si treatment.
2) Even if Al-5Mg thermal spraying is applied to the fluororesin paint coating, the thermal spraying has almost no effect on the weather resistance of the coating film.
3) The proposed method shows a higher weather resistance than the current method of partially coating on the thermal spray coating, which is easily deteriorated at an early stage.

## 5.Reference

[1] Kainuma S, Fujimoto H, Du J, Yang M, Muto K and Miyata H 2017 Fundamental study on corrosion resistance and anti-corrosion of interface between al-5mg thermal sprayed coating and heavy duty paint coating *J. Japan Soc. Civ. Eng. Ser. A1 (Structural Eng. Earthq. Eng.*

[2] Yang M, Muto K, Kainuma S, Du J and Miyata H 2019 Deterioration Characteristics of Overlapping Layers between Al-5mg Thermal Spray Coating and Heavy-Duty Paint Coating *Materials Science Forum*

[3] Kainuma S, Guo X, Kobayashi J, Muto K and Miyata H 2016 Fundamental study on anti-corrosion characteristics of al-5mg alloy thermal spraying coatings in highly corrosive environment with NaCl *J. Japan Soc. Civ. Eng. Ser. A1 (Structural Eng. Earthq. Eng.*

[4] Ochi S, Takayanagi T, Otsuka K, Shirato M, Ichiba M and Kainuma S 2018 Examination of hydrogen peroxide spray type accelerated weather resistance test (HP-Xe test) *Proceedings of JSCE Materials and Environments* (Tokyo) pp 141–2

[5] Wood K 2002 The effect of fluoropolymer architecture on the exterior weathering of coatings *Macromolecular Symposia*

[6] Mori K, Narita T, Tachi K and Umemura S 2001 A New Weatherability Test Method for Coatings with Acceleration Factor of 100 *J. Mater. Life Soc.*

# Analysis of the Structure of Heat Losses from Buildings and Benefits of Thermal Insulation Systems

**D Macek[1,*] and M Kosina[1]**

[1] Department of Construction Management and Economics, Faculty of Civil Engineering, Czech Technical University in Prague, Prague, Czech Republic

daniel.macek@fsv.cvut.cz

**Abstract.** The sources of energy are finite and are often found in unstable regions of the world, and we cannot assume at present that heat will be cheaper. Hence, and not only for these reasons, the issues of heat losses from buildings must be addressed. The paper investigates the issues of heat losses from buildings and analyses the benefits of thermal insulation of buildings. The thermal insulation of a building reduces the leakage of heat during the heating season and brings about savings in heat consumption amounting to tens of percent. In addition, when a building is insulated, the heating season may start later and end earlier. Typical thermal insulation technical modifications include the insulation of external walls, replacement of windows, roof insulation, insulation of the ceiling under the attic and above the cellar. Reducing the leakage of heat will reduce its consumption in heating and thus also cuts down air pollution with emissions produced by heating. The benefits of thermal insulation are in the economic, as well as technical sphere of building operation. The study compares the structure of heat losses and heat gains of apartment buildings from the 1970s, both in the insulated and uninsulated version, with houses complying with the current required standards and houses complying with the current recommended standards.

## 1. Introduction

The evaluation of heat leakage through external walls based on pre-set normative requirements only started in the early 1960s. These requirements were significantly tightened around 1980. Approximately 60% of all family houses in the Czech Republic had been built by the aforementioned year of 1980. In the early 1990s, another significant tightening of the requirements was introduced.

There are many ways of reducing operating costs. In the same way, in the case of economies in buildings, there are many ways of how to achieve them. The paper addresses one of them - thermal insulation. Thermal insulation systems cannot be perceived in a narrow spectrum as the way leading to economies, but globally: thermal insulation systems significantly affect other ways of making economies. One of them, which is very important, is the source of heating. In general, it can be said that a high-quality and well-prepared project and a subsequent insulation of the house will save 30-50% of heating costs. In low-energy to passive houses, the savings account for up to 70 - 97% of heating costs [1,2].

The paper focuses on the analysis of the development of the structure of heat losses from apartment buildings from the 1970s, in the insulated and uninsulated version, compared with houses complying with the currently required standards and houses complying with the currently recommended standards. Also, it summarizes the benefits of building insulation, both in the economic and technical sphere of building operation.

## 2. Problem analysis

A critical point of thermal insulation systems are thermal bridges. As the quality of thermal insulation of building envelopes, the fillings of openings, roofs and floors increases, the importance of removing thermal bridges also grows. A thermal bridge is a place in a structure with significantly worse heat technical properties than the rest of the structure. For this reason, there is an increased heat flow through it. More thermal energy leaks through it and it has a colder surface in the interior and, on the contrary, a warmer surface in the exterior than the surrounding structures [3,4]. This heat flow is illustrated in Figure 1. Critical points include the connection of the building envelope with foundations, the contact of fillings of openings with the building envelope and others. Thermal bridges are either of a linear (contact of two structures) or a point type (e.g. a dowel in the thermal insulation system). In addition, other undesirable accompanying phenomena occur in the places of thermal bridges, such as moisture condensation, the formation of mould or even damage to the structure itself [5,6,7].

**Figure 1.** Thermograph.

Table 1 display the heat losses/gains of a building from the 1970s and a today's building, both built in compliance with the applicable standards. The first row represents a typical house from the 1970s (**A**), while the second is the same house with additional thermal insulation (**B**). The third row characterizes a house of the same dimensions built in compliance with the current standard (**C**), and the last row represents a house built in compliance with the recommendations listed in the standard (**D**).

**Table 1.** The heat losses/gains.

| | thermal bridges | *windows* | *walls* | *ceiling* | floor | *ventilation* | internal heat gain | *solar gains* |
|---|---|---|---|---|---|---|---|---|
| **A** | 7% | 13% | 48% | 16% | 11% | 18% | -9% | -4% |
| **B** | 17% | 15% | 26% | 11% | 28% | 33% | -20% | -10% |
| **C** | 18% | 25% | 32% | 12% | 10% | 34% | -21% | -11% |
| **D** | 28% | 24% | 30% | 11% | 9% | 47% | -33% | -16% |

The effect of thermal bridges on the heat losses from a building presents the percentage of heat leakage through thermal bridges. If the heat losses from thermal bridges accounted for 7% of the heat loss and, after additional insulation, for 17% in the house from the 1970s, in houses built in compliance with the current standard they account for 18% in the case of required values and even 28% in the case of recommended values. This means that the heat losses through thermal bridges account for ¼ of the total heat loss from today's houses built in compliance with the standard recommendations. This is four times more than in the case of a house from the 1970s [8].

Windows, which accounted for 13% of the heat loss in the 1970s, currently account for 24% of the heat loss, which almost doubles the effect of windows. Heat losses through windows currently contribute to the heat loss by approximately the same percentage as the heat losses through thermal bridges. And additional insulation of a house from the 1970s hardly changes the contribution of windows to the heat loss.

Walls without additional insulation were the cause of almost one half of all heat losses in the 1970s. Additional insulation of these houses reduces the significance of heat losses through the walls to half the original value, i.e. to 26%. This means that after additional insulation of a house from the 1970s, the heat losses through the walls will only account for a quarter of the total heat loss. This value of heat losses through the walls roughly corresponds to the percentage of heat losses through the walls of houses built in compliance with the current standard where the heat losses through the walls participate in the total heat loss by only 2% more than thermal bridges.

In the case of the ceiling and the floor, no significant change in the contribution of these structures to the heat loss from the building has occurred since the 1970s. Historically, both structures follow the same trend as the walls where the heat loss through this structure accounts for a smaller part of the total heat loss. It is only in the case of additional insulation of a house from the 1970s that the percentage of heat losses through the floor doubles to a value of 28% as additional insulation of the floor only occurs in exceptional cases.

Ventilation accounted for 18% of the heat loss in the 1970s, which represents a value 2.5 times lower than the heat loss in a house built in compliance with the current standard for recommended values. This is caused by the values listed in the current standard which do not allow such high losses through other structures as in the 1970s. Therefore, ventilation currently accounts for a half of the heat loss from a house.

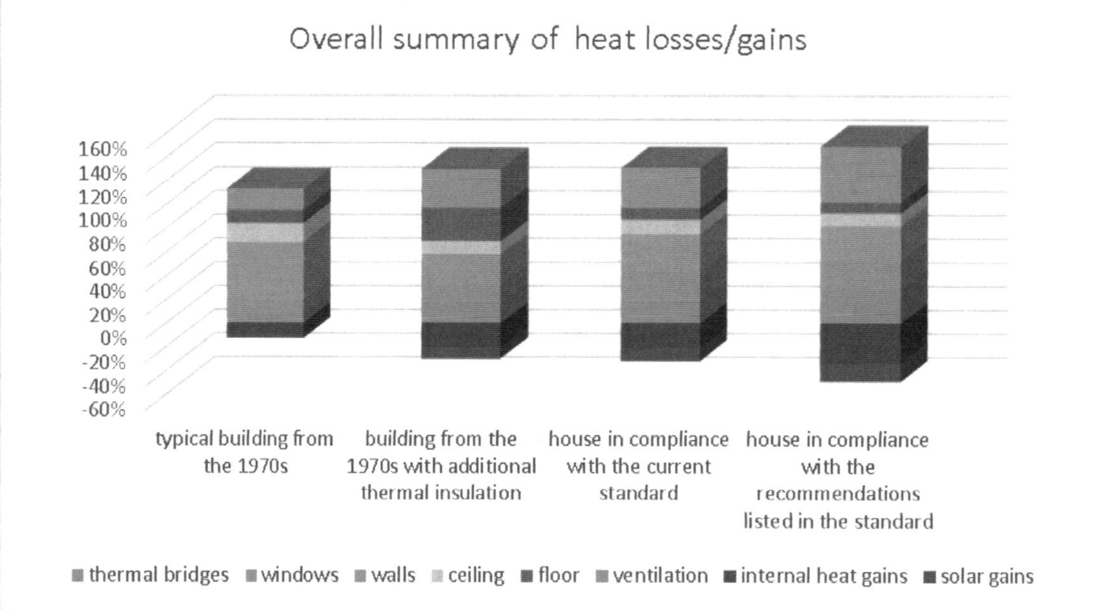


**Figure 2.** Overall summary of heat losses/gains.

The heat losses presented in Table 1 are shown as negative values, as these are internal heat gains and solar heat gains. According to the current standard, internal heat gains in a house amount to three times the value of a non-insulated house from the 1970s. The same applies to solar heat gains. Thus, Table 1 clearly shows the trend of the growing percentage of gains compared to the past.

Data from the Table 1 were used for the elaboration of Figure 2, which is the overall summary of the study. It shows that the biggest difference between the 1970s and the present is in the losses through thermal bridges. Their prevalence is due to better insulation of houses pursuant to the current standard where heat losses through weak points of the structure are manifested more significantly accounting for up to a quarter of the total heat loss.

Thermal bridges, windows and ventilation contribute to heat losses to a greater extent than in the 1970s. The opposite is true of walls, ceilings and floors. These structures participate in the heat loss from a house built in compliance with the current standard less than in the case of a house from the 1970s.

With additional insulation of a house from the 1970s (except the floor), the percentage of individual structures in the heat loss from the building is equal to their percentage in the house built in compliance with the current standard for the required values. This emphasizes the importance of thermal insulation; after its installation, the highest percentage of the heat loss from the building is represented by ventilation. Thanks to the reduction of heat losses by additional insulation of the building, the environmental impact resulting from its use is positively affected.

## 3. Conclusions

To conclude, the economic and technical benefits of thermal insulation of buildings can be summarized.

Economic benefits of insulation of buildings:

- reduced operating costs by improving the energy performance of the building,
- reduced energy consumption in heating - possibility of using a smaller and thus cheaper heating source,
- investments in thermal insulation will pay off the most at the time of a planned reconstruction,
- thermal insulation will increase the thermal comfort of the building allowing a more efficient use of the building.

Technical benefits of an insulated building:

- insulation may remove the condensation of water vapour on the interior surface of the building envelope which leads to the formation of mould, improved thermal comfort of rooms,
- less volatile temperature time patterns in structures (elimination of thermal bridges) and thus reduction of potential failures caused by temperature differences (lower thermal expansion),
- reduced overheating of the building in summer,
- reduced loading of the heating system - lower emissions,
- protection of the supporting outer shell of the building and improved acoustic properties.

Among other benefits of insulation, we can mention its positive effect on the structural performance of the house, improved masonry properties against the weather, improved acoustics, as well as improved health of the inhabitants – alleviation of allergies, rheumatic problems, cold related illnesses, and, last but not least, increased attractiveness of living in the building. After the installation of thermal insulation, the house will breathe better and so will indisputably its inhabitants.

**Acknowledgments**

This work was supported by the Grant Agency of the Czech Technical University in Prague, grant No. SGS20/015/OHK1/1T/11.

## 4. References

[1] Aberg M, Carlsson J and Nilsson A M 2018 Journal of Building Engineering 20 559-568
[2] Kisilewicz T, Fedorczak-Cisak M and Barkanyi T 2019 *Energy And Buildings* **205** 109541
[3] Kuan S, Chan C and Shu C 2014 *Int. Jour. of Smart Grid and Clean Energy* **3** 64-69
[4] Evola G, Costanzo V and Marietta L 2018 *Buildings* **8** 180
[5] O'Grady M, Lechowska A A and Harte Annette M 2017 *Applied Energy* **208** 1038-1052
[6] Counsell J M, Ameer A and Stewart M J 2016 *I. J. of Smart Grid and Clean Energy* **5** 229-236
[7] Najjar M K, Figueiredo K and Hammad, A W A 2019 *Jour. of Cleaner Production* **235** 789-800
[8] Yang L 2017 *Int. Jour. of Smart Grid and Clean Energy* **6** 182-189

# Application of Daylight System for Densely Populated Kampung in Jakarta: A Review

**T A Hanjani[1], O C Dewi[1,*], and N S D Putra[2]**

[1] Department of Architecture, Faculty of Engineering, Universitas Indonesia, Kampus Baru-Depok, Jawa Barat, 16424, Indonesia
[2] Department of Mechanical Engineering, Faculty of Engineering, Universitas Indonesia, Kampus Baru-Depok, Jawa Barat, 16424, Indonesia

*ova.candewi@ui.ac.id

**Abstract.** This study aims to review the special daylight system applied for housing situated in a densely populated kampung in Jakarta. Rapid urbanization has made housing in city areas even denser, where the high density makes the distance between buildings closer. The high density of area limits the vertical opening; thus, daylight cannot get through. The literature review is used as an alternative intervention by collecting, transmitting, and distributing literature to present the strength, weaknesses, and possible implementations of special daylight systems. This paper review that special daylight systems with required light pipes specifications in densely populated kampung can be used as alternative systems that can be applied to bring natural light into a densely populated village, with longer day duration in tropical climates. In addition to bringing light into densely populated villages, the special daylight system with light pipes can be further explored to see its relationship with users' comfort

## 1. Introduction

Densely populated areas in cities emerged due to rapid urbanization [1]. High-density areas make people live inadequate housing conditions, especially for the underprivileged [1], [2]. Jakarta, one of the most populous cities in Southeast Asia, [3] with an average population of 16,882/km2 and a tropical climate with a longer daytime duration of 11 hours 47 minutes - 12 hours 29 minutes [4] has the potential of making daylight as an alternative for sustainable energy. The high density of areas in Jakarta means that residing in densely populated villages lacks daylight [3] as it is one of the criteria for the adequate living condition [2], [5] and where artificial light is used throughout the day as a substitute. Poor daylight in figure 1 occurs due to the minimal distance between buildings caused by the opposite buildings' screening effects [1], [6], as shown in table 1.

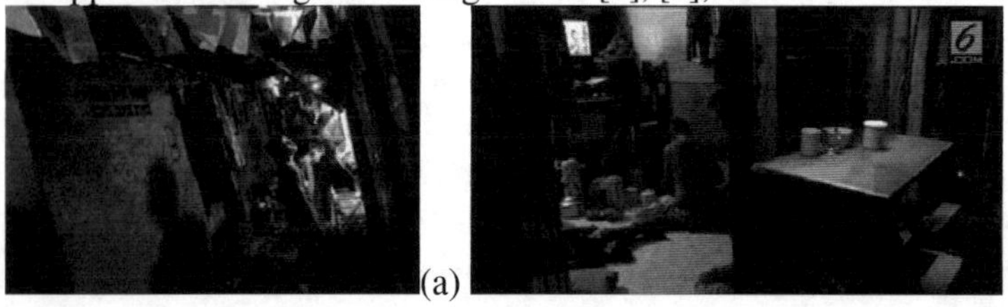

(a) (b)

**Figure 1.** Daylight conditions (a) at dense population village alley and (b) indoor throughout the day due to high population density in west Jakarta [7]

IOP Conf. Series: Materials Science and Engineering 1165 (2021) 012005 doi:10.1088/1757-899X/1165/1/012005

**Table 1.** Typical housing characteristic in dense population kampung in Jakarta

| | Characteristic |
|---|---|
| Floor area | Range from less than ten sqm with an average of 21 sqm up to 36 sqm and three up to four people reside together [8]–[10]. |
| Building height | Stacked with lower floor ceiling located just below the floor level above, mostly low ceiling height [10]–[12]. |
| Building material | Roof finished in wood beam cover in corrugated roof some in concrete. The floor is covered with ceramic tile over the concrete floor, sometimes bare concrete floor[10] and concrete or bricks used for wall material. |
| Opening | It is difficult to do since it continuously extended to deal with space limitation[10] and attached to each other [12]. |

People living in densely populated villages cannot access daylight because daylight is very much influenced by climate, hence orientation [13]–[16], windows [13], [17], spatial shape [13], [14], [17], [18], room divider [13], colour [13], [14] and visibility [13], [14], [19] are factors affecting special daylight systems as shown in table 1. Daylight in residence meets the standard if it gets sufficient daylight from 08.00 to 16.00 or at least 8 hours a day and has a light level [20], according to the room's function range from 60 lux up to 300 lux and the activities carried out [21]. This study aims to review the daylight system through literature review by collecting, transmitting, and distributing literature to seek alternative systems that can be applied to bring natural light into a densely populated village, as shown in figure 2.

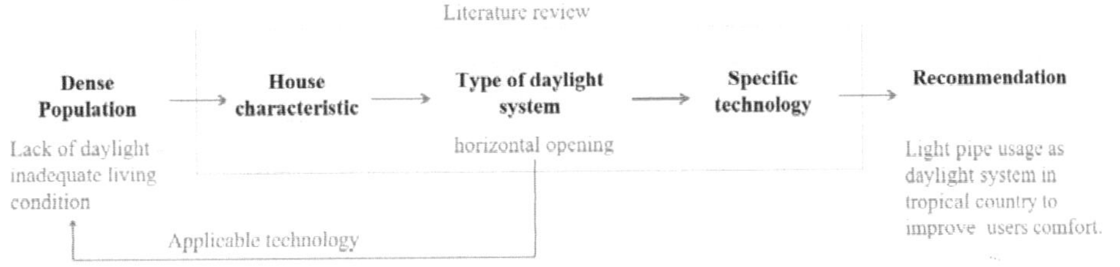


**Figure 2.** Review framework.

## 2. Types of daylight system used in densely populated areas

Limited space in densely populated kampung characteristic areas makes windows or vertical openings impossible for building façades [22]. Therefore special daylight systems with horizontal surfaces [13], [19], as in table 2, are used to bring light into housing at densely populated kampung in given limited space without changing housing characteristics.

**Table 2.** Types of special daylight system on the upper side of the building

| Type | Description | Strength | Weakness | Possibilities of implementation, adjustment, and barrier |
|---|---|---|---|---|
| Lightwell or light shaft | Top atrium without roof connects space[23] based on a window to width ratio[13]. | The inner wall uses a reflective material, such as a mirror. The size of the hole can be reduced when using a mirror. | The intensity of the light distribution is directly proportional to the size ratio. | Impossible to be implemented with adjusted to floor height, given the limited space [23]. |
| Tubular skylight | Duct-like tube [13] to capture, transport light into building through the tube | Transmits 50% daylight Effective when used for one store occupancy | Light intensity is directly proportional to the window ratio Less effective than lightwell | Impossible to be implemented only for multi-floor [13], [25]. |

IOP Conf. Series: Materials Science and Engineering 1165 (2021) 012005 doi:10.1088/1757-899X/1165/1/012005

| | | | | |
|---|---|---|---|---|
| | [24]. | Economical<br>Suitable for use in tropical areas with high sun elevation. | The transmitted light depends on the size and length of the tube | |
| Skylight with dynamic mirror | An active skylight transmits light into space through moving components like a reflector, mirror, or mechanical device [13], [24]. | It follows the sun's movement; therefore, there is more sunlight even at low sun angles.<br>Used in the basement. | Use the built-in mirror to reflect light<br>Use a PV electric motor to drive the mirror<br>Expensive | Impossible to be implemented because it requires high maintenance. |
| Beam daylight | Heliostat with mirror mounted-sun track device reflects light through the upper part of the building [13], [26]. | It uses the mirror on the heliostat as a sun detector to reflect light at any position.<br>Used in the basement or deep plan building. | Using lenses and mirrors<br>Depends on the substrate surface | Impossible to be implemented because it requires many mirrors and a good building structure. |
| Light pipe or light tube | Roof lighting with high reflective pipe or tube to direct light [27], [28]; therefore, it is also called tubular daylight guide system[22]. | Light pipes are made of hollow pipes.<br>Pipe material can be prismatic, aluminum alloy, silver coat, and other similar material types.<br>Reflects light with internal reflection.<br>Multi Floor<br>High level of efficiency | Utilizing internal reflection so that external factors do not influence the reflection level depends on the pipe material's reflectance factor.<br>Because it is straight, mirrors are used to bend light.<br>The lighting level decreases when it is too long or bent. | Possible to be implemented because it does not take up much space and can be used for multi-floor with several apertures and spesification[29]. |
| Prismatic system | Prismatic glass tile above window [13] for directional control[14]. | Popular because it can refract light through a triangular surface that works similarly to a prism. | Expensive<br>The angle between the light guide and the prism must be positioned below 90' to transmit it. | Impossible to be implemented when angles are required given the limited space. |
| Glass floor | Glass paving block to transmit light at lower floor[13]. | Used in the basement area<br>Able to transmit light from one floor to another.<br>The small size makes damage will not be immediately visible. | Paving blocks look alike.<br>Sturdy but easily broken | Impossible to be implemented when there are many objects on the floor because it will block the transmitted light. |

Table 2 shows the type of special daylight system used on the upper side of the building with its strength and weakness, leading to the conclusion of its possibilities to be implemented in typical houses in a densely populated kampung in Jakarta. Out of seven types, only light pipes are possible to be implemented. The ability of a light pipe to collect, transmit and distribute light through a reflection in the pipe is affected by the surface of the pipe plane as a reflectant [26], [30], geographic position, Daylight Factor (DF), Daylight Autonomy (DA), CIE sky condition, and aspect ratio affect the light pipe's ability to capture and transmit light [22], [31]. Besides, the quality of daylight also depends on the window to floor ratio, the shape and location of the light opening, reflectant factor from the surface in the room, and the position of the light opening against external factors that can block the entry of light. Light intensity is determined by the size of the optimal opening adjusted for the height of the building [23]

## 3. Analysis and discussion

Previous studies have shown that light pipes function optimally in warmer climates and low latitudes because they are exposed to more sun, where the transmission of light can be improved by using light pipe specification; understanding architectural feature and orientation [6], [22], [23], [29], [30], [32]–[35]. Light pipes collect, transmit and distribute daylight in the space through reflections in pipes [26] in spaces with limited access to daylight [27], [29], [32], such as densely populated kampung. Daylight systems with light pipes are used as an alternative to be developed in densely populated kampung of Jakarta, located at the equator and have a longer daytime duration than sub-tropical countries. Although light pipes use internal reflection to transmit daylight, light pipe produces higher light transmission than tubular guides. The degree of lightness and light distribution is directly proportional to the window size. The light pipe specifications, such as diameter, length, and aperture used, determine the light that enters the space. Therefore, the light pipe needs to meet several specifications as in table 3.

The light pipe's luminance level is affected by sky conditions, Commission Internationale de l'Eclairage (CIE) overcast or sunny, and the aspect ratio. Darula stated that 1) If the length of the light pipe is short, the inner surface of the pipe must have a high reflectance level; 2) The high amount of reflection in the light pipe can result in low lighting levels, so the entry angle of the natural light must be high, and; 3) if the light pipe is long, then the diameter needs to be adjusted so that the lighting level meets the standard. The diameter is directly proportional to the level of light[32]. The larger the diameter, the less light reflection on the pipe resulting in bringing the light deeper [31], [32], [35], [36]. This condition makes light pipes widely used for single-story buildings because if they are used for multiple floors, a pipe connection is required on each floor, resulting in lower lighting levels and higher costs [29].

**Table 3.** Light pipe specification and its characteristic [29], [31], [35]

| Specifications | Characteristics |
|---|---|
| Ratio | The lighting level depends on the pipe height level.<br>The shorter the pipe is, the less light is lost.<br>The diameter or length of the pipe affects the stray light when transmitted. |
| Pipe | Lightweight material contributes to less structure change. |
| Optical components | A short light pipe needs shade to avoid glare.<br>Long light pipe: requires an optical element to direct light.<br>Material: has a high transmission value. |
| Collector | Glazing material with high transmission value.<br>Low maintenance level: resistant to pollution. |
| Diffuser | Needed to avoid glare, efficient on sunny sky but not very efficient during overcast conditions<br>Diffusers with a transmittance value of 50% offer more benefit in CIE overcast conditions and sunny skies. |

| | |
|---|---|
| Interior light pipe | The level of reflectance in the pipe determines the light pipe's lighting level.<br>The reflectance level used is 0.98 or 0.93 |
| Light pipe design | Collector orientation: placed in the south part of the building to obtain radiation throughout the day<br>High light pipe efficiency when the light-receiving part (collector) is slightly bent (has an angle).<br>When the collector is tilt to an angle of 6 'in a straight and wide underside -wide near the bottom- it produces more energy.<br>In sub-tropical climates, light pipes with angled geometric shapes receive more light than light pipes with standard (straight) geometric shapes.<br>Light can directly enter the pipe because the sun is directly above the equator. |
| Geographical position | Tropical climate - high sun exposure, higher light intensity than sub-tropical climate. |
| Sky condition and solar radiation | The tubular light guide is efficient for direct solar radiation and can transmit light when overcast conditions. |

## 4. Future work recommendation

Studies on special daylight systems with light pipes are related to light distribution and energy efficiency in sub-tropical countries. Therefore, a future study related to daylight can be developed further to look at the application of daylight systems with light pipes in densely populated areas in tropical climates and their impact on users' comfort. This paper seeks to reference light pipes development in tropical countries and increase users' comfort.

## Acknowledgment

This research funded by PUTI-UI Saintekes grant research under contract no:NKB2439/UN2.RST/HKP.05.00/2020. The authors would like to thank the Department of Architecture, Faculty of Engineering, Universitas Indonesia.

## 5. References

[1] J. Adianto, A. Okabe, E. Ellisa, and N. Shima, "*The Tenure Security and Its Implication to Self-Help Housing Improvements in the Urban Kampong: The Case of Kampong Cikini, Jakarta,*" Urban Reg. Plan. Rev., vol. 3, no. 0, pp. 50–65, 2016, doi: 10.14398/urpr.3.50.

[2] UN-Habitat, "*Sustainable urbanization in Asia A Sourcebook for Local Government,*" 2016. [Online]. Available: www.unhabitat.org.

[3] Kompasiana, "*Kawasan Terpadat di Asia Tenggara dan Rawan Kebakaran di Jakarta* Halaman all - Kompasiana," *2019.*

[4] S. Wirjohamidjojo and Y. Swarinoto, "*IKLIM KAWASAN INDONESIA (Dari Aspek Dinamik - Sinoptik),*" 2010.

[5] Kementrian Pekerjaan Umum dan Perumahan Rakyat, *Permen PUPR No. 10/PRT/M/2019 tentang Pedoman SMKK, no. 1.* 2019, pp. 1–25.

[6] R. Ménard and J. Souviron, "*Passive solar heating through glazing: The limits and potential for climate change mitigation in the European building stock,*" Energy Build., vol. 228, p. 110400, 2020, doi: 10.1016/j.enbuild.2020.110400.

[7] "FOTO_ *Minim Cahaya Matahari, Warga Gang Venus Nyalakan Lampu 24 Jam Penuh* - Page 2 - News Liputan6." .

[8] Kementrian Pekerjaan Umum dan Perumahan Rakyat, *Keputusan Menteri Permukiman dan Prasarana Wilayah No: 403/KPTS/M/2002.* 2004, pp. 1–5.

[9] R. Suryantini and T. Panjaitan, "*Domestic Energy Practice in Vertical Housing,*" Int. J. Built Environ. Sci. Res., vol. 01, no. 02, pp. 79–90, 2017, doi: 10.24853/ijbesr.1.2.79-90.

[10] E. Ellisa, "*Coping with crowding in high-density kampung housing of Jakarta,*" Archnet-IJAR, vol. 10, no. 1, pp. 195–212, 2016, doi: 10.26687/archnet-ijar.v10i1.790.
[11] T. Amemiya, A. Okabe, A. Suzuki, E. Ellisa, and A. Hery Fuad, "*Megacity Skeleton,*" Japan Archit. Rev., vol. 2, no. 4, pp. 426–434, 2019, doi: 10.1002/2475-8876.12110.
[12] F. Lirenzsa, E. Ellisa, and A. Paramitha, "*Spatial negotiation in domestic space of the home-based garment industry in Kampung Tambora, Jakarta,*" IOP Conf. Ser. Earth Environ. Sci., vol. 447, no. March, p. 012034, 2020, doi: 10.1088/1755-1315/447/1/012034.
[13] N. Lechner, *Heating, Cooling, Lighting Sustainable Methods for Architects*, 4th ed. John Wiley & Sons, Inc., 2015.
[14] G. Gordon, *Interior Lighting for Designer*, 5th ed. New Jersey: John Wiley & Sons, Inc., 2014.
[15] H. T. I. So, "Light Shapes Space: How Light Influences Our Perception of Space," Ryerson University, 2010.
[16] A. Omishore, P. Mohelník, and D. Míček, "*Light Pipe Comparative Study,*" Sel. Sci. Pap. - J. Civ. Eng., vol. 14, no. 1, pp. 17–26, 2020, doi: 10.1515/sspjce-2019-0002.
[17] G. Lippsmeier, *Tropenbau Building in the Tropics*. München: Callwey, 1980.
[18] O. H. Koenigsberger, "*Manual of tropical housing and building.*" Longman, London], p. v., 1974, [Online]. Available: file://catalog.hathitrust.org/Record/007473523.
[19] M. Azizkhani and J. Haberl, "*Assessment and discussion of the level of the application of passive/natural systems and daylighting systems by practitioners in the US,*" Sci. Technol. Built Environ., vol. 4731, 2020, doi: 10.1080/23744731.2020.1783961.
[20] SNI-6197-2011, "*SNI-6197-2011 tentang Konservasi Energi Pada Sistem Pencahayaan,*" 2011.
[21] SNI 03-2396-2001, "*Tata cara perancangan sistem pencahayaan alami pada bangunan gedung,*" 2001.
[22] M. Sibley and A. Peña-García, "*Flat glass or crystal dome aperture? A year-long comparative analysis of the performance of light pipes in real residential settings and climatic conditions,*" Sustain., vol. 12, no. 9, 2020, doi: 10.3390/su12093858.
[23] A. alah Ahadi, M. R. Saghafi, and M. Tahbaz, "*The optimization of lightwells with integrating daylight and stack natural ventilation systems in deep-plan residential buildings: A case study of Tehran,*" J. Build. Eng., vol. 18, no. March, pp. 220–244, 2018, doi: 10.1016/j.jobe.2018.03.016.
[24] F. Sharp, D. Lindsey, J. Dols, and J. Coker, "*The use and environmental impact of daylighting,*" J. Clean. Prod., vol. 85, pp. 462–471, 2014, doi: 10.1016/j.jclepro.2014.03.092.
[25] O. Šikula, J. Mohelníková, and J. Plášek, "*Thermal CFD analysis of tubular light guides,*" Energies, vol. 6, no. 12, pp. 6304–6321, 2013, doi: 10.3390/en6126304.
[26] I. Ullah, "*Heliostats daylighting system for multi-floor buildings,*" J. Daylighting, vol. 6, no. 2, pp. 202–209, 2019, doi: 10.15627/jd.2019.18.
[27] S. Darula, R. Kittler, and M. Kocifaj, "*Luminous effectiveness of tubular light-guides in tropics,*" Appl. Energy, vol. 87, no. 11, pp. 3460–3466, 2010, doi: 10.1016/j.apenergy.2010.05.006.
[28] D. Phillips and C. Gardner, *Daylighting: Natural light in architecture*. Massachusetts: Architectural Press An Imprint of Elsevier, 2012.
[29] A. Goharian and M. Mahdavinejad, "*A Novel Approach to Multi-Apertures and Multi-Aspects Ratio Light Pipe,*" J. Daylighting, vol. 7, no. 2, pp. 186–200, 2020, doi: 10.15627/jd.2020.17.
[30] H. Von Wachenfelt, V. Vakouli, P. Diéguez, and N. Gentile, "*Lighting Energy Saving with Light Pipe in Farm Animal Production,*" vol. 2, pp. 21–31, 2015, doi: 10.15627/jd.2015.5.
[31] G. E. Toledo, A. V. Pelegrini, and A. Heemann, "*Design parameters for solar light pipes in the Brazilian context,*" Int. J. Sustain. Eng., vol. 9, no. 4, pp. 251–258, 2016, doi: 10.1080/19397038.2016.1148796.
[32] S. Darula, M. Kocifaj, and J. Mohelníková, "*Hollow light guide efficiency and illuminance distribution on the light-tube base under overcast and clear sky conditions,*" Opt. - Int. J. Light Electron Opt., 2013, doi: 10.1016/j.ijleo.2012.09.052.
[33] A. P. Diéguez, N. Gentile, H. Von Wachenfelt, and M. Dubois, "*Daylight Utilization with Light Pipe in Farm Animal Production : A Simulation Approach,*" vol. 1, pp. 1–11, 2016, doi:

10.15627/jd.2016.1.

[34] M. Hamdani, S. M. E. A. Bekkouche, T. Benouaz, R. Belarbi, and M. K. Cherier, "*Minimization of Indoor Temperatures and Total Solar Insolation By Optimizing the Building Orientation in Hot Climate,*" Eng. Struct. Technol., vol. 6, no. 3, pp. 131–149, 2014, doi: 10.3846/2029882x.2014.988756.

[35] L. Sharma, S. F. Ali, and D. Rakshit, "*Performance Evaluation of a Top Lighting Light-Pipe in Building and Estimating Energy Saving Potential,*" Energy Build., 2018, doi: 10.1016/j.enbuild.2018.09.022.

[36] K. Vasilakopoulou, A. Synnefa, D. Kolokotsa, T. Karlessi, and M. Santamouris, "*Performance prediction and design optimization of an integrated light pipe and artificial lighting system,*" Int. J. Sustain. Energy, vol. 35, no. 7, pp. 675–685, 2014, doi: 10.1080/14786451.2014.932281.

IOP Conf. Series: Materials Science and Engineering 1165 (2021) 012006 doi:10.1088/1757-899X/1165/1/012006

# GIS-based Architectural Design in Heritage Landscapes: towards a Knowledge-based Definition of Touristic Routes

**M López Sánchez[1,*] and M Linares Gómez del Pulgar[1]**

[1] Instituto Universitario de Arquitectura y Ciencias de la Construcción, Universidad de Sevilla, Av. Reina Mercedes 2, 41012, Seville, Spain

marinalopez@us.es

**Abstract.** Network-based spatial analysis tools are useful for solving complex routing problems. However, the search for the most optimal route in terms of distance and time does not always guarantee the most convenient design solution, as occurs in heritage landscapes. A landscape architecture project aimed at designing touristic routes that connect heritage resources requires additional variables as in these cases it is not only a matter of achieving the most efficient connection between the heritage assets, but of detecting the paths that allow a better interpretation of the cultural meaning of the landscape. This text presents a GIS-based analysis protocol that combines least-cost path network analysis with weighted overlay analysis in order to introduce the time depth, the visibility and the level of visual and noise pollution of the paths as variables of influence in the selection of the touristic routes. The proposed method has been applied in the metropolitan area of the city of Seville (Andalusia, Spain), where there are relevant heritage resources that remain unconnected and could benefit each other through integrative touristic formulas. The aim of the method is advancing a protocol for a computer aided advanced architectural design in territorial scales.

## 1. Introduction

The dominant ways of thinking and designing in architecture have substantially changed in the last decades, confirming the need of advancing novel formulas for achieving more scientific decisions. Bringing science into the design process without compromising the creativity is a mayor challenge in architectural research.

In the path towards an informed design, the access to geographic data during the design process has been acknowledged as helpful, especially when architectural design engages with territorial scales. Building a strong foundation of georeferenced data through Geographical Information Systems (GIS) extends our analytical capabilities and allows a better understanding of the territorial context we are dealing with, as well as of the impacts of our interventions on it [1]. The GIS has an inherent integration potential, as it is capable of functioning as both a tool and a language shared between different spheres of knowledge and agents [2]. For this reason, it allows the integration of multiple layers of territorial information as history, land use, hydrology, meteorology, topography, etc. in the design process.

This paper presents a GIS-based method for reaching scientific decisions in landscape architecture design. It particularly deals with the task of selecting paths for implementing a touristic project in a heritage landscape. Touristic routes, the object of this research, differ from cultural itineraries. The first ones are contemporary products created for touristic convenience [3][4], while cultural itineraries have been originated by historical processes that have turned them into paths with heritage value

[5][6]. While the identification of cultural itineraries responds to historical research, the definition of touristic routes requires analysis in terms of their potential both as networking and communication tools. They are specifically aimed at structuring the heritage offer of a cultural destination [7] as well at offering a clear reading of the cultural landscape they cross by equipping the walking experience with minor architectures that run along them and that facilitate the heritage interpretation process through viewpoints or scenic stops with interpretive panels, etc.

For this reason, it is not just about detecting the most efficient route between the heritage assets but of identifying the paths that allow a better interpretation of the cultural meaning of the landscape. This requires variables as time depth, visibility and visual/noise pollution to be considered. The trails which cross the landscape areas with the highest heritage value and that present fewer disturbing elements and wider viewsheds will allow a better appreciation and knowledge of the cultural landscape.

## 2. Materials and methods

The proposed route design method combines GIS least-cost path network analysis with weighted overlay analysis in order to introduce the time depth, the visibility and the level of visual and noise pollution of the paths as variables of influence in the selection of the touristic routes. The least-cost path analysis identifies the most optimal routes in terms of distance and time for connecting the heritage resources. The weighted overlay analysis starts by turning diverse base layers into value layers that distinct land areas according to different assigned levels, as described in table 1.

**Table 1.** Criteria and values for reclassifying the base layers

| | Base layer | Criteria | Value |
|---|---|---|---|
| Time depth | Land use change map | Areas with the same land use since the first land use record | 5 |
| | | Areas with minor land use change since the first land use record (e.g. change of type of cultivation in agricultural fields) | 3 |
| | | Areas with mayor land use change since the first land use record (e.g. from agricultural use to industrial use) | 0 |
| Visibility | Viewsheds of the most relevant heritage items and cultural landscape features. | Areas where is possible to observe more than 10 heritage items or landscape features | 5 |
| | | Areas where is possible to observe between 5-10 heritage items or landscape features | 3 |
| | | Areas where is possible to observe less than 5 heritage items or landscape features | 0 |
| Noise pollution | Noise pollution map | No noise pollution recorded in the area | 5 |
| | | 55-60 dB(A) | 3 |
| | | >60 dB(A) | 0 |
| Visual pollution | Visual pollution map witch identifies visual disturbing elements in the area | Areas out of the buffer zones of the disturbing elements | 5 |
| | | Areas between 200 and 400 meters far from any disturbing element | 3 |
| | | Areas less than 200 meters far from any disturbing element | 0 |

The criteria have been designed according to the case study is meant to serve, but they can be modified to different densities of heritage items, landscape features and/or disturbing elements. Once all the base layers have been reclassified into the three defined values, a final layer of five levels of land suitability for touristic routes is obtained from a weighted sum that assigns 40% of influence for time depth, 30% for visual pollution, 20% for visibility values, and 10% for noise pollution. The contrasted observation between the land suitability map and the optimal routes obtained by network analysis allows the designer to reach a final decision which balances time and distance effectiveness with scenic, landscape and heritage values. The described method is adjustable so it could evolve progressively by the addition of more variables and value ranges.

## 3. Results and discussion

The proposed method has been applied in the zone of the metropolitan area of the city of Seville (Andalusia, Spain) shown in figure 1, which is characterized by a high density of heritage resources. This area of approximately 100 $km^2$ covers a very wide temporal arc as it includes chalcolithic dolmens, the archaeological remains of El Carambolo, a protohistoric Phoenician temple, the archaeological remains of the Roman city of Italica, the monastery of San Isidoro del Campo (XIV century), the monastery of la Cartuja (XV-XVI centuries) and interesting vernacular architecture of modern and contemporary times like the hacienda de Torrijos or Divina Pastora. This heritage richness allows to trace the evolution of human settlement in the territory from the first occupations to the present with a complementary and differentiated approach from that offered by the city.

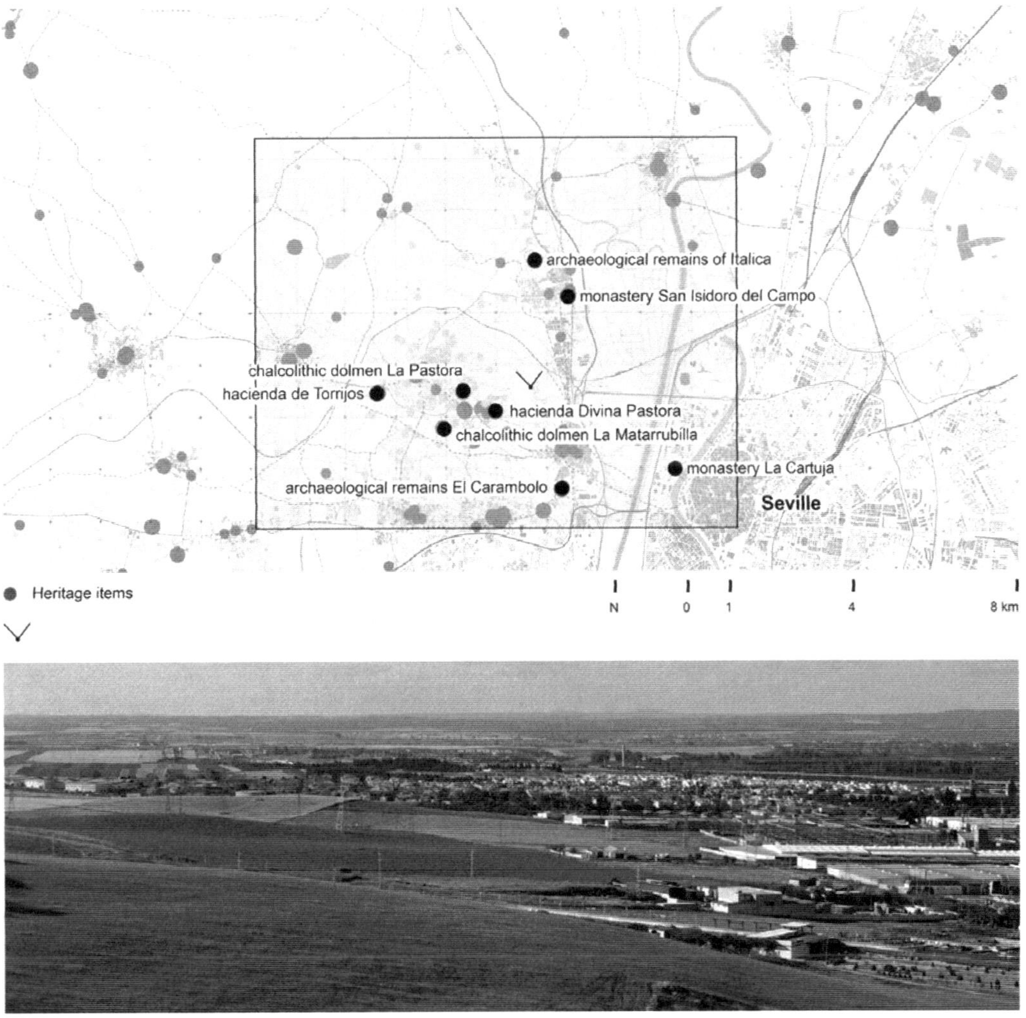


**Figure 1.** Case study area. Source: map by M López and photography by A Tejedor (2019)

However, these heritage items do not establish relationships each other or with the landscape and, consequently, the historical connections between them remain hidden from the population. It is a representative example of those periurban areas where the landscape diversity and the time depth of the territory coexist with fast roads, industrial areas or large commercial infrastructures that fragment the landscape and greatly hinder its heritage reading.

The methodology has been applied to this area using the software ArcGIS® y ArcMap™ 10.4.1. from Esri. It can be observed how the studied variables, as shown in figure 2, allow to reach a final route design more committed with the particular landscape values of the area.

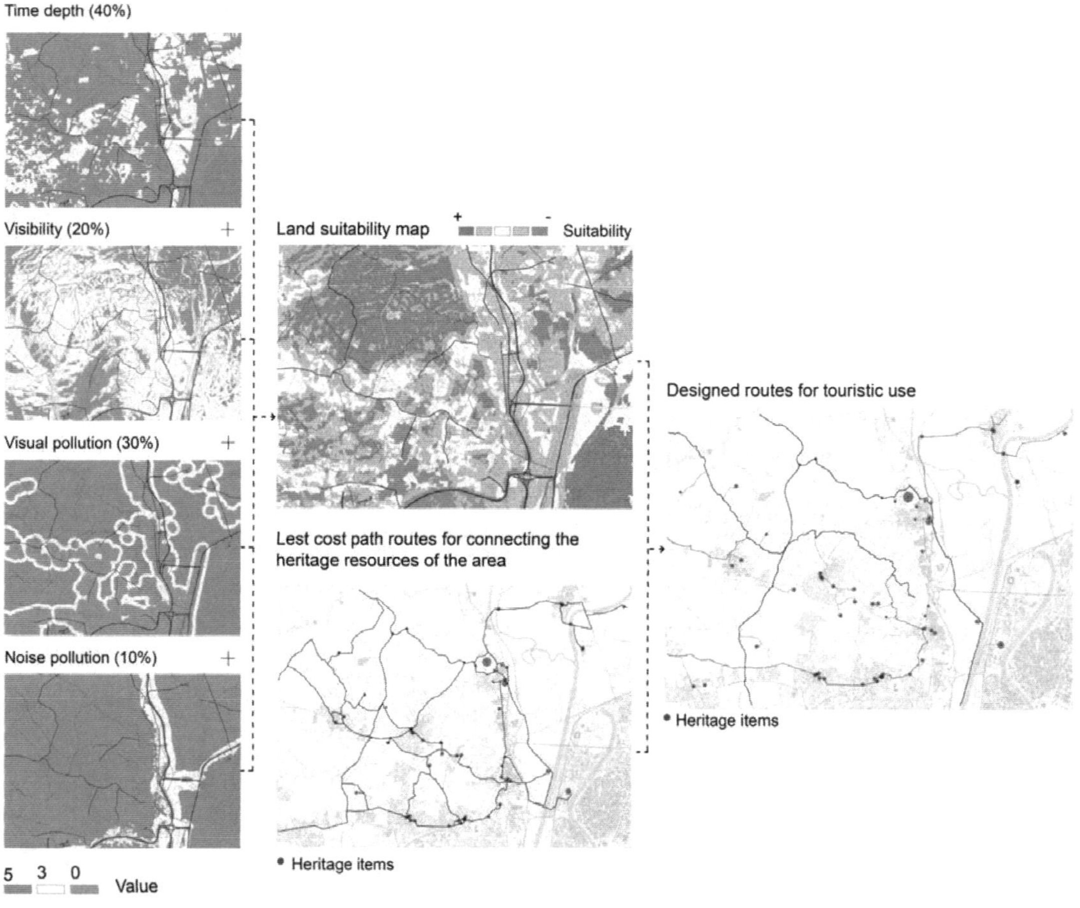


**Figure 2.** Results of the analysis. Source: M López

## 4. Conclusion

The sustainable development of our territories demands a greater relation between spatial planning and design with landscape and heritage [8]. In times of emergency such as the current one, the need to move forward into informed site-specific designs becomes vital. The described method combines a standardized quantitative formula, which is network analysis, with a weighted overlay system whose variables, criteria and weights can vary depending on the particular characteristics of the territory under study as well as the particular view of the designer. In periurban areas, measuring pollution levels or attending to land use change patterns becomes essential, but different contexts could need distinct parameters. The benefit of the method is to offer possibilities for re-thinking and adjustment, which are key for design-related contexts.

In the paths towards an informed architectural design, methods that are committed with the particular components and dynamics of a territory have to be developed. On the other hand, as architecture would never be a purely scientific discipline since creativity is equally attached to it as technique, methods should also be capable of working as a stablished system with a fixed protocol that, however, allows adaptations to different views and progressive modifications through the design process.

**Acknowledgments**
This research is funded by the regional government of Andalusia (Junta de Andalucía) with the support of The European Regional Development Fund thorough the project "GISPARQ. Intelligent and Sustainable Management of ARQuitectonic Heritage: digital innovation and integrated management for conservation, tourism and the economy of culture" (US-1263780). Main author would also like to thank the Spanish Ministry of Education, Culture and Sport for financial support via the research grant: "Formación del Profesorado Universitario" (FPU16/02768).

## 5. Reference

[1] Steinitz C 2012 *A Framework for Geodesign: Changing Geography by Design* (Redlands: ESRI Press)
[2] López Sánchez M, Tejedor Cabrera A and Linares Gómez del Pulgar M 2020 Arquitectura y Sistemas de Información Geográfica: Hacia un Proyecto de Paisaje Informado *Proyecto, Progreso, Arquitectura* **22** pp 72–87. http://dx.doi.org/10.12795/ppa.2020.i22.04
[3] Timothy D J and Boyd S W 2014 *Tourism and trails: Cultural, ecological and management issues* (Bristol: Channel View Publications)
[4] Ward Perkins D, Beckmann C and Ellis J 2019 *Tourism Routes and Trails: Theory to Practice* (Wallingford: CABI)
[5] Arcila Garrido M, López Sánchez J.A. and Fernández Enríquez A 2015 Rutas turísticos-culturales e itinerarios culturales como productos turísticos: reflexiones sobre una metodología para su diseño y evaluación. *Análisis espacial y representación geográfica: innovación y aplicación* ed de la Riva J, Ibarra P, Montorio R and Rodrigues M. (Zaragoza: Universidad de Zaragoza-AGE) pp 463–471
[6] Hernández Ramírez J. 2011 Los caminos del Patrimonio. Rutas turísticas e itinerarios culturales *Pasos* **9(2)** pp 225–236. https://doi.org/10.25145/j.pasos.2011.09.021
[7] Moulin C and Boniface P 2001 Routeing heritage for tourism: Making heritage and cultural tourism networks for socio-economic development. *International Journal of Heritage Studies* **7(3)** pp 237–248. https://doi.org/10.1080/13527250120079411
[8] López Sánchez M, Tejedor Cabrera A and Linares Gómez del Pulgar M 2020 Guidelines from the Heritage Field for the Integration of Landscape and Heritage Planning: A Systematic Literature Review. *Landscape and Urban Planning* **204** 103931. https://doi.org/10.1016/j.landurbplan.2020.103931

# Building Envelope as to Improve Home Quality in Microhome Design Approaches

**R E Diana[1] and F Oktarina[1*]**

[1]Department of Architecture, Art & Design Faculty, Universitas Multimedia Nusantara, Scientia Boulevard, Gading Serpong, Tangerang, Indonesia

freta.oktarina@umn.ac.id

**Abstract.** The cities around the world face challenges with the shortage of land and increased demand for housing. Limit the land use for area development might resolve urban land scarcity problems. In this case, the efficiency of land use and microhome has seen as space-saving solutions. The research discusses *home* term on micro-scale through spatial concept design approaches. The study uses the qualitative method that interpreted by various schematic drawings. Each approach will affect the building envelopes as well as increase the home quality. The finding shows that the building envelope might be seen as inseparable attributes, consist of one or more elements; roof, wall and window, secondary skin or floor, functions as complementary attributes or a direct response to the context. This paper is an initial study to generate microhome spatial quality by using the building envelope.

## 1. Introduction

Avoiding the development of built environments over the green area and controlling new development areas are parts of the sustainable development approaches. In the Spatial Planning Guidelines, no. 26 the year 2007 mentioned that city spatial planning must contain 30% area for greenspaces of the total city area. The total area of the green spaces in the big cities in Indonesia, specifically Jakarta, Bogor, Depok, Tangerang, and Bekasi (Jabodetabek), has hugely decreased. The decrease over the last 30 years reaches 60 thousand hectares per year. Lately, the total area of the green spaces in Jabodetabek decreased from 30% to 10%[1].

However, Indonesia's demographic shows a rapidly growing youth population that leads to increased housing demand. The Programme of Action is a result of the International Conference on Population and Development (ICPD) in 1994 aimed to reach a sustainable development program by controlling lower population growth. It shows an intimate relationship between demographics and sustainable development. In the 1950-2100 periods, young age fertility equal to the natality rate in Indonesia is the highest among South-East Asian countries. The national medium-term development plan RJPMN *(Rencana Pembangunan Jangka Menengah Nasional)* programmed to decrease the natality rate that would simultaneously increase the youth population[2]. Following this program, the governmental development sector must be improved, including the needs of housing. Restrain the new development and limit land development is the key to resolve decreased green spaces. Limited space delivers microhome concepts as a space-saving solution.

Based on Stoneham's (2015) argument; *'What makes a house a home?' It is not something that can be universally answered, but without a thorough understanding of the occupants' psychological nature, a home cannot be achieved. Therefore, an understanding of the human condition, regardless of whether it is the individual or the collective, must preface any residential architectural design. Once this understanding is developed, the architecture would have a significant impact on the mind*[3].

Objected to the "*home*" term, the research based on microhome design project that was chosen as a topic study in the early even semester in Multimedia Nusantara University's 2019/2020 academic year[4]. The project focus on exploring micro "*home*" design concept. Extracted from the project, the authors research the exploration of the spatial concept as initiated by activities study.

## 2. Methods

The author extracted conceptual approaches from the entire design process, mainly the schematic design phase. Indeed, architectural space ideas analysed with the schematic drawing, clustered into several aspects by grouping methods as if used in Concept Sourcebook that describe five sets of concept vocabularies; functional grouping and zoning, architectural space, circulation and building form, response to context and building envelope[5]. In this research, the list divided as follows to identify the general group:

### *2.1. Spatial Concepts*

Spatial concepts as a most used approach to the home design consist of spatial relations and spatial qualities. Spatial relation concept, both indoor-indoor and indoor-outdoor relation analysed with orthogonal projection through section drawings. Meanwhile, the spatial qualities refer to the attributes of each room interpreted by detail drawing.

### *2.2. Building Envelope*

The building envelope was reviewed as the spatial concepts' results, interpreted on orthogonal projection through elevation drawings.

At the schematic design stage, the main concepts defined as the interpretation of both forms and spatial order. Through this research, the authors compile the defined ideas in the sorts of specified schematic drawings:

- Massing and the roof forms via perspective drawing
- Spatial relations via plan and section drawing
- Space qualities via section drawing and interior detailing
- Building envelope and openings via elevation drawing

## 3. Results and Discussion

Design process highly depends on the designer's values, attitudes, life views and general behaviour patterns[6]. Designing is a psychological consideration that differs over the way to see the problems. The designer's perceptions about the project will be some of the most critical thinking in the planning process. Conceptual process in the *home* term might initiate from user needs and activities study and its relation to spatial terms.

Heidegger's opinion, that he suggested: *"building" and "dwelling" has proper relation, named activities that had once been – and, for him, should become again – fundamentally inseparable, conjoined in a "thinking" susceptible to immediacies and enormities of daily existence*[7]. The essential human condition explained as all about the activities which serves the needs and wants of contemplation in a living body. The Active and Contemplative term closely related to the definition, the character, and the way of human life[8]. Both of the terms manifested in the architectural project state as human activities; the daily routines and its meaning in spatial scale. Those conceptions projected by gaining the specific activities to derive the main concept as a basic rule to set criteria and boundaries.

Concepts in architecture are normally thought of as belonging to the schematic design stage of the planning process. This traditionally has been where the designer has generated the "big ideas" for the building design[5]. The concept is the primary framework for most of the architecture project stages within programming, schematic design, design development, working drawing or construction stages.

The authors study the schematic drawing used as primary data to review, collect, analyse, and interpret various schematic concepts. Using the Concept Sourcebook[5] as the vocabulary to read the design process, the studies look further into the spatial concepts as a strategy for designers to detail architectural concepts. These strategies are selected and developed refers to relevancies and connectivity between one or more design aspects in a project. The designer defines the *home* through a variety of conceptual approaches. Refers to the study, there are three sets of approaches which explained as follows:

### *3.1. Spatial Relations Approach*

The designer sees the room to room interrelationship as the right response to link the activities inside and surround the home. Spatial relation consists of indoor spatial relations and indoor-outdoor spatial relations: plan and section drawing used as the main tools to interpret the main concepts. As the designer's priorities, spatial relation concepts indicate functional zoning, user and the activities interrelationship, spatial adjacencies, spatial organization, and openings placement. Zoning tends to classify the place horizontally or vertically. At the first step, the designer focus on indoor spatial relations. Plan drawing used to interpret the main concepts. The plan is projected into the section drawing to show the vertical relationship between the stories. The building envelope tends to be complementary attributes as if only the plan extruded. Most design used the kind of mono-pitch roof with the number of roof planes depends on the combined mass number—the examples of the conceptual design interpreted in figure 1.

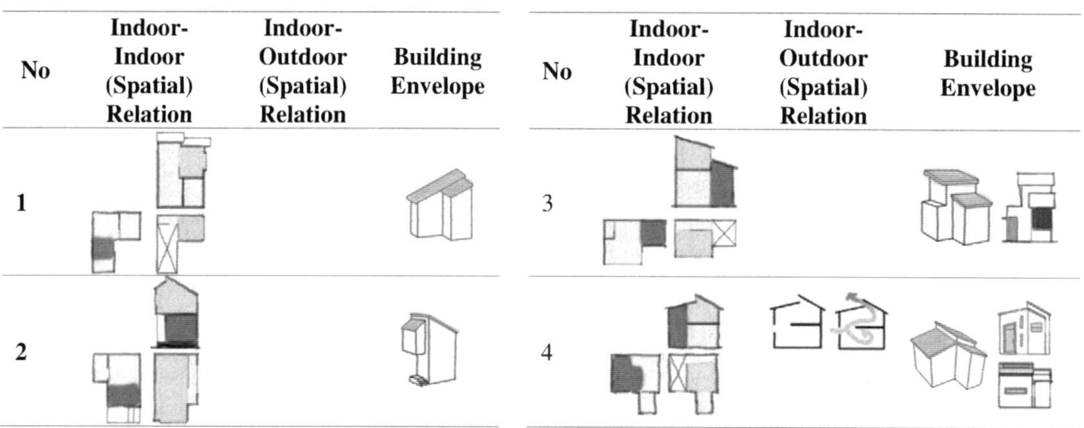


**Figure 1.** Interpretation of Indoor Spatial Relations Approaches *(Analysis, 2020)*

### *3.2. Response to context approach*

Besides the indoor spatial relations, the designer focuses on the site inventories as one of the main issues that responded through the building envelopes. The site inventories could be the property boundaries, tree, sunlight, water, existing building, traffic system, and approach to the building. The building envelope considered to respond to one or more items to improve home quality. The building envelope might conceptualize as below:

- The roof unites the whole mass,
- The air circulation, lighting and view may be accessed through the openings,
- The wall is conceptualized as an open-able area to widen the inside floor.

Figure 3 shows the conceptual design interpretation.

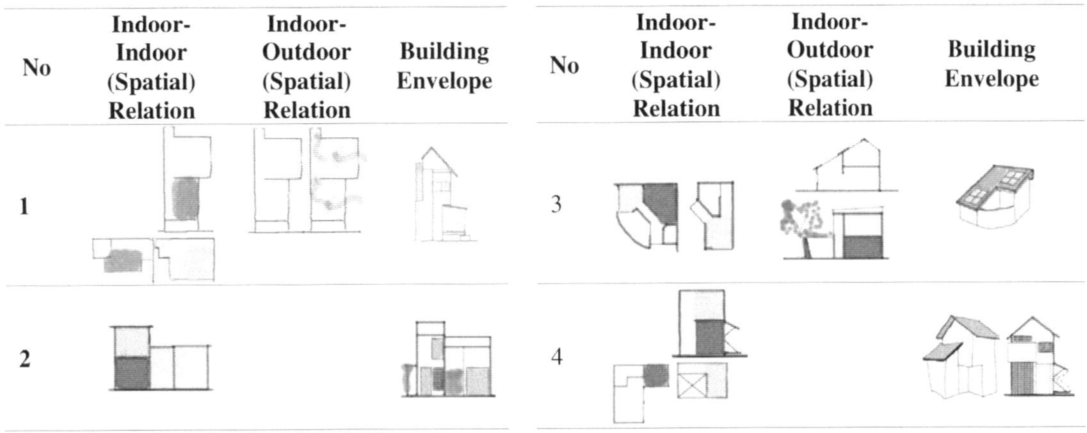


**Figure 2.** Interpretation of Response to Context Approaches *(Analysis, 2020)*

*3.3. Spatial qualities approach*

The designer programs the required environments as responses to certain activities. Room by room quality is an important thing to reach the design goals. The designer specifies the qualities through various ways such as accommodated activities, interior detailing, lighting, need for adjacencies, circulation types or wall, floor and ceiling concepts. Spatial relations developed to link the conceptualized rooms. The figure below shows the conceptual interpretation of this approach.

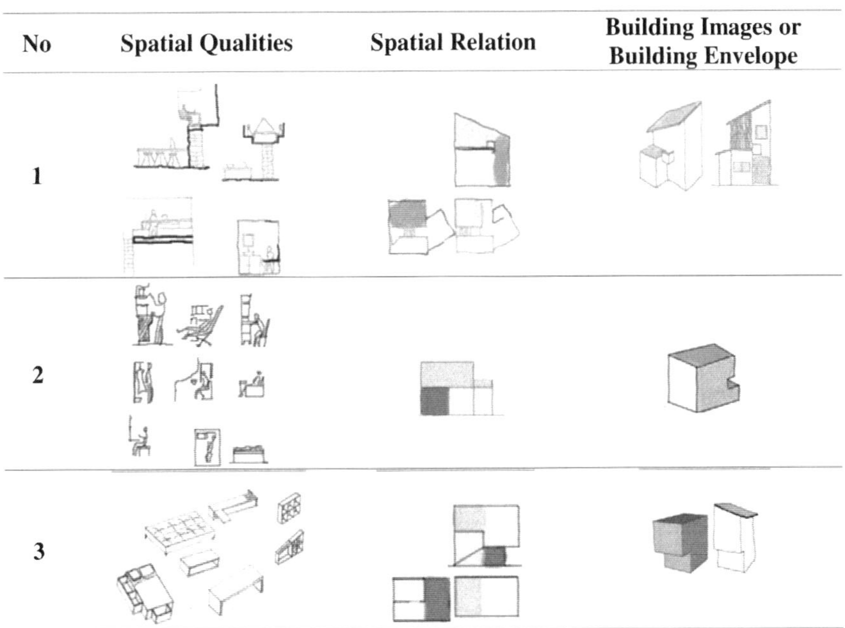


**Figure 3.** Interpretation of Spatial Qualities Approaches *(Analysis, 2020)*

## 4. Conclusions and Recommendation

The *home* term might define a variety of conceptual approaches. Refers to the microhome design project state, the spatial concepts composed of the spatial relations, respond to context, and spatial qualities might use to approach microhome design. Those approach selected influences the conceptual process and the architectural forms, especially the building envelope.

In developing the spatial concept, the building envelope might be seen as inseparable attributes to define home quality. The building envelope might consist of one or more elements; roof, wall and window, secondary skin or floor, functions as complementary attributes or a direct response to the context. This research might be an initial study to improve microhome design through the building envelope. By designing various kind and pattern of the building envelope like the opening, it might improve home quality as well. As if design the opening would particularly affect microhome lighting and ventilation systems. The interpretation as follows in figure 4.

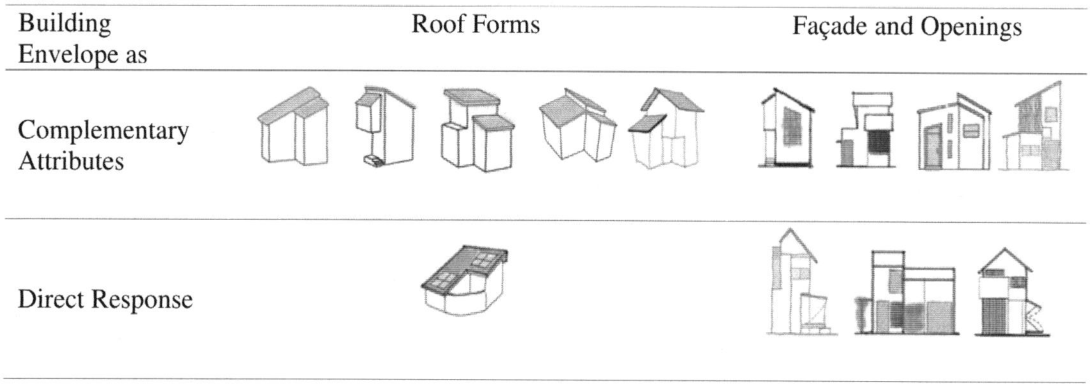


**Figure 4.** The Building Envelope in Microhome Design Project *(Analysis, 2020)*

**Acknowledgments**
The authors would like to thank Diah Esti Sihanani (lecturer), the coordinator of AR322 Architecture Studio class who arranges the microhome design project, Jonathan Danilo Alvar, Amalia Bella Sinthia and Vira Savira Putri Shea (students) who assisted in collecting and interpreting data.

## 5. References

[1] http://www.biomaterial.lipi.go.id/menyelamatkan-ruang-terbuka-hijau-kota/, accessed on October 14th 2019
[2] Rajagukguk W & Samosir OB 2015 *Dinamika Demografis Indonesia 1950 – 2100* (Jakarta: National Population and Family Planning Commission)
[3] Stoneham B & Smith D 2015 *The House and the Home: The Balance of Architecture and Psychology within the Residential Home"* (Geelong: Deakin University)
[4] https://elearning.umn.ac.id/course/view.php?id=170, accessed on May 30th 2020
[5] White E T 1975 *Concept Sourcebook: A Vocabulary of Architectural Forms* (Arizona: Architectural Media Ltd)
[6] White E T 1972 *Introduction to Architectural Programming* (Arizona: Architectural Media Ltd)
[7] Sharr A 2006 *Heidegger's Hut* (London: The MIT Press)
[8] Arendt H 1958 *The Human Condition 2nd Ed* (Chicago: The University of Chicago Press)

IOP Conf. Series: Materials Science and Engineering 1165 (2021) 012008 doi:10.1088/1757-899X/1165/1/012008

# A Building Information Modelling (BIM) Approach to the Systematic Management of Construction Projects

**Mahmoud Ershadi*[1], Marcus Jefferies[1], Peter Davis[1], and Mohammad Mojtahedi[2]**

1. School of Architecture and Built Environment, University of Newcastle, Newcastle, Australia
2. Faculty of Built Environment, University of New South Wales, Sydney, Australia

Mahmoud.ershadi@uon.edu.au

**Abstract.** Recent advancements in information technology have benefited the construction sector in more effective storage, analysis, and transmission of information. Making proper project decisions is hinged upon the verified and timely information provided by systems. It is especially the case in large construction projects in which many partners collaborate to deliver a project. Building information modeling (BIM) as an intelligent 3D model-based approach provides construction managers with insight on how to integrate elements of project configuration. In this paper, we theorize the potential contribution of BIM to the systematic management of projects at the project and corporate level. The findings guide practitioners on how to take advantage of this approach towards the integrated planning and control of construction projects.

## 1. Introduction

The process of creating a three-dimensional digital model of a building, which enables the visualization and integration of all its components is called Building Information Modelling (BIM) [1]. This method disseminates the application of computer-aided design in the construction industry, which has integrated the application of information systems and analytical software packages towards better decision-making, scheduling, estimating, and maintenance of construction projects [2]. Across different construction functions and outcomes, it is of utmost importance to utilize BIM for boosting performance and effectiveness [1]. Theoretical developments in BIM suggest its benefits to geometric modelling of a building's performance [3].

The emergence of BIM concepts in scholarly literature dates back to the early 2000s when it was used to deal with overlaps in virtual design, construction, and facility management [4]. Some studies attempted to describe its contribution to the project management process since then, among which we can refer to Rokooei [2] who described the correspondence of BIM and project managers' roles. Further, Bryde et al [3] highlighted the benefits of utilizing BIM based on a cross-section of construction projects. Previous research on project management implications of BIM underlines its role as an integrator of workflows and corresponding outcomes [2], which enables coordinated design, production, communication, and data analysis [4]. This study takes one step further and makes a theoretical discussion on how BIM integrates design components of construction projects and also interact with other corporate systems such as finance, maintenance, and document control towards systematic project management.

## 2. BIM contribution at the project level

According to the Project Management Body of Knowledge (PMBOK) [5], ten areas represent standard project management activities. Project integration management as the first domain holds the overall project together and brings outcomes together. The object-based nature of BIM enables the proper configuration of documents and products. One of the major concerns in this process includes integrated change control (ICC) [6] that can be significantly facilitated by the application of BIM via communication of change requests, accumulation of data for analyzing side effects, and incorporation of the perspectives of key project decision-makers on specific changes. The second process includes project scope management that defines project boundaries and ensures compliance of work with a scope baseline [5]. BIM can contribute to minimizing deviations from the approved scope by maintaining visual features of ultimate deliverables (buildings, facilities, etc) and comparing the progress against respective technical/design requirements. The systematic view of BIM on how work packages are being delivered based on the work breakdown structure (WBS) allows project managers to trace deviations upfront before scope creep occurs.

The third area of project management refers to project schedule management describing the process of defining and timing tasks. Software packages that are used in BIM help visualizing and animating the time-based sequence of activities. The technique is known as 4-dimensional project scheduling which adds the time factor to the 3D architectural design. The benefits of this feature include detecting logical issues and clashes, as well as improving the effectiveness of activity sequencing [7]. The BIM analytical outputs also provide information for setting clear and attainable milestones in the planning phase. The application of four-dimensional (4D) scheduling has further extended to include cost estimates, which is the subject of the fourth project management domain [5]. Five-dimensional planning in BIM allows project managers to highlight the costs allocated to activities through the progression of time [8]. This feature integrates design with cost estimates that provides project managers with insight on the structure of project costing and the relationship between cash flow and the sequence of work.

The fifth area refers to managing the quality requirements as promised to clients and stakeholders. The contribution of BIM in this area is to identify clashes and conflicts of building design [3]. It significantly helps designers to improve the quality of their design in terms of technical or even aesthetic characteristics. Lou et al. [9] introduced a quality control platform based on BIM referring to the centralization of quality specifications, real-time data sharing, and comparative analysis of quality data. Project resource management also can be benefited from BIM by linking the five-dimensional (5D) model and resource loads. It illustrates how resources are being allocated among activities in relation to different parts of the building, providing reliable insight into optimizing their distribution among different areas in the progression time. Stakeholder management and communications management are the other two interrelated domains aiming to manage the human side of projects by applying a variety of methods such as interpersonal skills, conflict resolutions, and communication skills [5]. BIM enhances collaborations both internally, among members of project teams, and externally with clients and partners so that productivity is maximized and multi-party decisions are accelerated.

Project risk management is another important domain aiming to safeguard the project against hazards and to control the impact of potential events (both positive and negative) on objectives [5]. BIM enables the identification of high-risk areas that may cause issues when design/engineering features are being implemented on-site. Further, the simulation outcomes guide practitioners to visually review and handle constructability issues associated with all elements of design [2]. Specifications can be accordingly improved by designers to mitigate the risk of improper implementation of engineering/architectural designs during construction stages. The other domain includes procurement management that identifies outside procurement needs and also plans for meeting them. Procurement decisions rely on the correct estimation of resources. In this regard, the process of estimating quantity take-off (QTO) in terms of necessary labor and material can be facilitated by the utilization of BIM modules.

## 3. BIM contribution at the corporate level

The contribution of BIM to systematic project management is not limited to operational aspects but also extends to corporate levels through accumulating and linking data from various operational parts of a project with corporate systems so that functional teams that contribute to project outcomes can benefit from BIM outputs. It is key to promoting the organizational learning process. This also constructs a basis for continuous improvement of key functions such as engineering, architectural design, scheduling, estimating, maintenance, and operation. This paper highlights the most important links between BIM and corporate systems as follows. Nevertheless, their proper integration under a broader platform lies with system architects.

### *3.1. The link with project management information systems (PMIS)*

PMIS incorporates several software applications that constitute a platform for collecting and utilizing project information across an enterprise [10]. Time and cost as the two most important elements of construction projects should be realistically estimated and established to form baselines against which performance is measured. The 4D and 5D BIM enrich the project schedule and cost baseline. Such outputs help senior managers in making informed decisions and, thus, should be closely linked with PMIS to enable comparing the performance against realistic plans. More importantly, whenever changes are made to project scope, BIM simulation should be reattempted and updates imported to the PMIS. BIM also complements PMIS to form an effective means for collaborative management and better communication of change orders, approved changes, non-compliance reports, and work inspection requests, which encourage higher transparency and accountability.

### *3.2. The link with financial information systems (FIS)*

The cash flow forecast is a fundamental component of financial planning. A realistic timing of earnings and expenditures from projects constitutes a backbone of robust long-term financial plans. 5D BIM gives clues to construction engineers to verifies cost estimates and optimize quantity takeoffs, which accordingly contribute to the development of more reliable cash flows based on a reliable estimation of labor, material, and equipment. This output of BIM needs to be linked to financial management systems in order to help financial planners accumulate project costs for drawing on corporate-level cash-outs.

### *3.3. The link with knowledge management systems (KMS)*

One of the benefits of BIM application is clash detection which uncovers errors in the engineering/architectural design process. Hence, valuable knowledge is being created in this process to help designers amend the specifications, procedures, and applied techniques. Lessons learned in terms of design issues should be captured, maintained, and shared with other projects, which implies the extraordinary importance of linking BIM outputs with knowledge management systems.

### *3.4. The link with document management systems (DMS)*

Document management systems allow predesignated users to store or retrieve project documents to form a shared database. BIM has a special package for document management which needs to be integrated with the corporate document management systems to be used for future correspondence and to be used by functional managers (as per defined access). This interrelationship contributes to centrally controlling project documents such as contracts, masterplans, charter, designs, issue logs, communications in a construction firm.

### *3.5. The link with maintenance management systems (MMS)*

Six dimensional BIM (6D BIM) links attribute data associated with product lifecycle (such as installation information, optimal operation conditions, maintenance requirements, and lifespan) with graphical information to understand the cost of assets during the lifespan. Such information is required to support facility management and operation. Thus, 6D BIM output needs to be properly incorporated into the corporate maintenance management systems so that regular maintenance and optimal operation can be adequately planned for achieving optimal conditions of assets.

### *3.6. The link with safety management systems (SMS)*
Zhang et al [11] argued that BIM outputs are useful for automatic safety checking of construction models. Visual analysis of the construction models gives insight into detecting safety hazards, especially fall-related risks, which need preventive measures to be in place before entering the construction phase. Thus, establishing a link between BIM and safety analysis software and systems enables developing more comprehensive construction safety method statements.

## 4. Systematic project management approach through BIM
Adopting a systematic approach in project management mitigates the risks of issues during construction by simulating the design, verifying estimates and schedules, and visualizing the possible outcomes before starting the construction work. To achieve systematic project management, the integration of data and workflows are of crucial significance at the project level in connection with corporate systems. BIM as a simulator and integrator plays a key role as the main enabler for collaboration in managing the project lifecycle stages. The physical and functional aspects of the building design are quantitatively analyzed concerning planning parameters. The outcomes are shared among stakeholders on time in order to speed up communications and avoid any misunderstanding or conflicts on configurations. Figure 1 map out the contribution of BIM to other systems (corporate level) and processes (project level) towards taking advantage of the digital representation of the project.

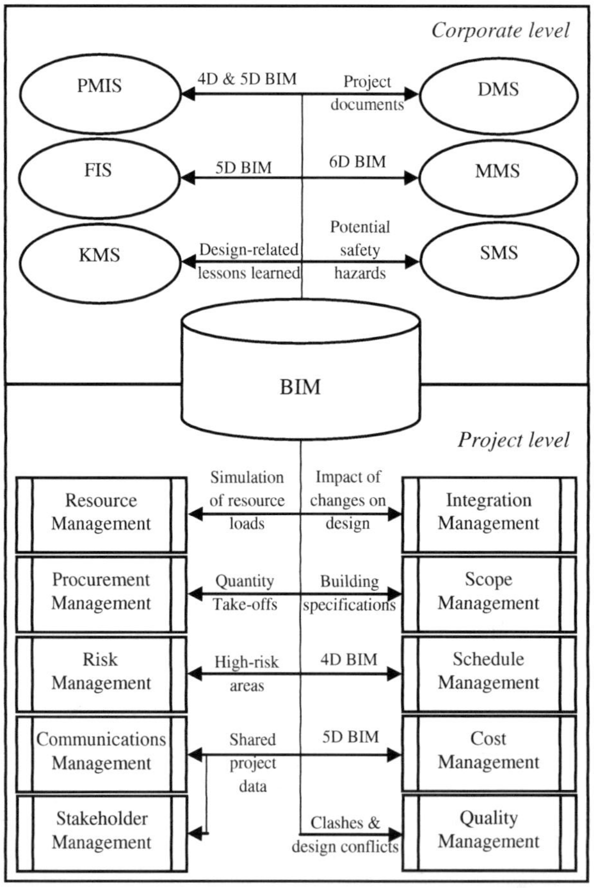


**Figure 1.** Systematic project management through BIM.

## 5. Conclusions
This paper aims to improve the understanding of systematic project management through BIM in the construction sector. We theorized that the interaction of BIM should be considered not only at the

project level but also at the corporate level. We conclude that at the project level, BIM contributes to mitigating various risks through simulating the project based on the 3D model and planning parameters. The benefits of BIM in managing each aspect of key project management processes were highlighted that asserts the positive role of BIM in verifying the planning parameters concerning design features. Animating the sequence of work in 4D is one of the most valuable outcomes allowing project planners to modify and improve timings as well as the overall logic.

In the corporate level, BIM outcomes should be properly integrated with corporate systems to promote improvements in the engineering and design process. Six systems were identified to be closely interrelated with BIM outcomes, implying the pivotal role of BIM in improving engineering and design processes, as well as enriching records for better operation and maintenance of assets. The results contribute to a safer environment since potential high-risk areas can be recognized using visual simulations. Finally, the authors suggest further research elaborates on each contribution of BIM to explain how relevant processes can be improved.

## 6. References

[1] Jung, Y. and M. Joo, 2011. Building information modelling (BIM) framework for practical implementation. Automation in construction, 20(2): p. 126-133.

[2] Rokooei, S., 2015. Building information modeling in project management: necessities, challenges and outcomes. Procedia-Social and Behavioral Sciences, 210: p. 87-95.

[3] Bryde, D., M. Broquetas, and J.M. Volm, 2013. The project benefits of building information modelling (BIM). International journal of project management, 31(7): p. 971-980.

[4] Harris, J., 2013. Integration of BIM and business strategy. Evanston, USA: Northwestern University.

[5] Project Management Institute, U., 2017, A Guide to the Project Management Body of Knowledge (PMBOK® Guide) Sixth edition. Project Management Institute: USA.

[6] Thiry, M., 2002. Combining value and project management into an effective programme management model. International journal of project management, 20(3): p. 221-227.

[7] Turkan, Y., et al., 2012. Automated progress tracking using 4D schedule and 3D sensing technologies. Automation in construction, 22: p. 414-421.

[8] Smith, P. 2015, Project cost management with 5D BIM, Proceedings of the 29th IPMA world congress.

[9] Lou, J., J. Xu, and K. Wang, 2017. Study on construction quality control of urban complex project based on BIM. Procedia engineering, 174: p. 668-676.

[10] Belaya, A.I., I.V. Tchancev, and M.A. Kossukhina. 2017. The selection of the corporate information system for innovative project management tasks. IEEE Conference of Russian Young Researchers in Electrical and Electronic Engineering (EIConRus).

[11] Zhang, S., et al., 2013. Building information modeling (BIM) and safety: Automatic safety checking of construction models and schedules. Automation in Construction, 29, p. 183-195.

# Numerical Simulation of the Stress Corrosion Rupture Zone for Deep Tunnels

**Xin Zhang[1], Jinshan Sun[2,*], Min Lang[1], Yuxiang Du[2], Suzhi Zhao[1], Qian Dong[2]**

[1] China Railway 18 Bureau Group Co.,Ltd, Tianjin 300222, China.

[2] Hubei Key Laboratory of Blasting Engineering, Jianghan University, Wuhan 430056, China.

sunjinshan@cug.edu.cn

**Abstract.** Stress corrosion is a geological phenomenon that damages the long-term stability of underground projects. A comprehensive approach for stress corrosion prediction of underground projects combining creep experiment, particle flow code (PFC)and Finite element method (FEM) is proposed. Calibration of the numerical model was undertaken according to the data obtained from laboratory tests. Based on the numerical model, biaxial and triaxial stress corrosion compression tests were conducted to investigate the stress corrosion mechanism of Jinping marble. Through numerical tests, the stress corrosion process was visually observed. According to the development history of microcracks, the stress corrosion rupture process of surrounding rocks can be divided into 4 stages, i.e., rapid growth, slow growth, rapid growth and instant leap. The process was controlled by driving stress ratio and confining stress. On or near the tunnel wall, most stress corrosion microcracks are induced by tensile stress and the direction is parallel to the direction of the maximum principle stress. Near the yield zone of tunnel, the ratio of shear microcracks becomes larger with increasing confining stress. And the amount of microcracks is larger at the rupture stage than that on the tunnel wall, which makes the stress corrosion rupture process look like a ductile failure process. There is a linear relationship between the logarithm of stress corrosion lifetime and driving stress ratio if the confining stress keeps constant. Stress corrosion lifetime will increase when $p_c$ becomes higher or driving stress ratio becomes lower. Based on the laboratory and numerical tests, the stress corrosion rupture depth of the tunnel section was predicted using FEM, which can optimize the length of bolts.

## 1. Introduction

The stress corrosion behavior of rock often leads to geological hazards or engineering diseases, and it includes stress corrosion damage and stress corrosion rupture, which affects the long-term stability of rock projects, such as tunnels and slopes. However, in the stress corrosion process of rock, the initiation and growth of cracks on the surface can be observed directly, but the internal microcracks are difficult to observe. In rock engineering, understanding the stress corrosion mechanisms of rock is important for the long-term stability of projects.

In the research of stress corrosion behavior of rock, Holcomb and Martin, and Chandler [4,10] studied the Lac du Bonnet granite. Géraldine [13] studied clay rock by uniaxial compression experiments. Cao and Xian [1] studied coal rock experimentally, and analyzed its micro-mechanical mechanism. Zhao [20] studied limestone experimentally, and introduced it into the numerical models. Liu et al. [9] studied the Jinping marble experimentally and numerically. Chen [3] studied the stress corrosion rupture characteristics of sandstone experimentally.

Many scholars have studied the constitutive models of stress corrosion damage and the damage evolution equations [2,15,16,17,19]. In these studies, the long-term strength of rock or rock mass has been studied. Experimental data show that the long-term strength of rock mass is about 62% ~ 75% of its transient strength [7]. Huang [13] inferred that the long-term strength of rock in uniaxial compression is about 77% of its transient strength by adopting the renormalization group theory.

The above researches mainly focus on the stress corrosion behavior of rock or rock mass on the macroscopic scale. In the past few years, researchers have been conducted on the microscopic characteristics of rock with discrete element method(DEM), such as Particle Flow Code (PFC2D/ PFC3D) [5,6]. Potyondy [11,12] proposed a parallel-bonded stress corrosion model (PSC) to simulate the stress corrosion failure process of rock in PFC2D and PFC3D. Using this model, Sun et al. and Liu et al. [8,14] analyzed the stress corrosion process of Jinping marble.

The remainder of this paper is organized as follows. In Section 2, the stress corrosion damage of surrounding rocks of Jinping II hydropower station is introduced. In Section 3, the test data is used to calibrate the micro-mechanical parameters of PFC specimens and the rational and feasible numerical simulation method is established. In Section 4, the validated numerical model is used to study the microscopic mechanism of stress corrosion mechanism of Jinping marble and to predict the final strength parameters. In Section 5, the empirical formula of long-term strength of rock is established, and the rupture zones of the tunnel are simulated by FEM. Finally, conclusions are given in Section 6.

## 2. Stress corrosion behaviors in Jinping II hydropower station

The Jinping II hydropower station is located in Sichuan Province, China. It includes four diversion tunnels, two auxiliary tunnels, and one drainage tunnel. Each diversion tunnel is approximately 16.67 km, and they are the deepest tunnels in China, with a typical depth of 1500 m~2000 m and the deepest depth of 2525 m. The maximum geostress measured in the field is 113.87 MPa.

The surrounding rock of the tunnels is mainly brittle marble. The average uniaxial compression strength of the rock is about 65~90 MPa, and the tensile strength is about 3~6 MPa. We performed 7 sets of uniaxial compression tests, 9 sets of triaxial compression tests, and 49 uniaxial rheological compression tests on Φ50mm rock specimens. The uniaxial compression strength of marble is 95 MPa to 107 MPa, the average strength is 97.7 MPa, the elastic modulus was 51 GPa, and the Poisson's ratio was 0.2.

In the excavation process of 4 tunnels, the phenomenon of spalling or rupture appears when the depth is more than 1500m. As illustrated in Figure 1 and Figure 2, spalling often start in 3 to 5 hours after excavation, and serious rupture occurs at the section of 50 ~ 60m away from the excavation face.

**Figure 1.** Spalling on the tunnel wall excavated by TBM

**Figure 2.** Creep rupture on the tunnel wall excavated by blasting

To measure the damage degree of the surrounding rock mass, the ultrasonic testing at several sections was carried out. It is considered that the rock mass has been damaged substantially when the ultrasonic velocity in the rock mass is reduced by 10%. For example, at 15+250m section of 3# tunnel whose buried depth is about 1750m, the depth of damage of rock mass increased remarkably after 6 months, as illustrated in Figure 3.

Moreover, the acoustic velocity of rock mass in the damaged zone decreases obviously after several months, which indicates that the damage degree has gradually increased. For example, the acoustic velocity of the damaged rock mass in the depth of 1.9m decreased about 16.7% after 5 months.

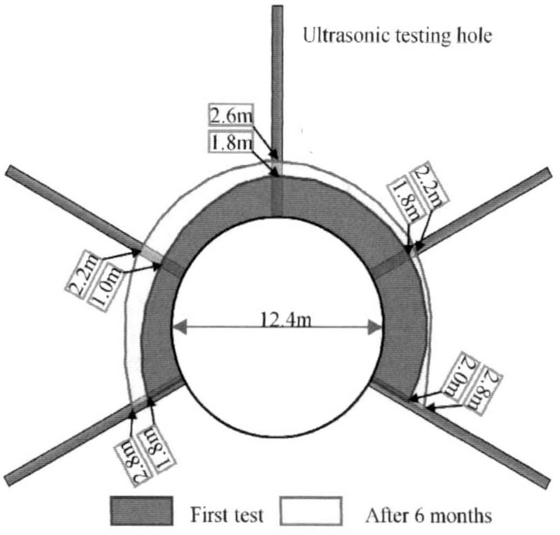


**Figure 3.** Ultrasonic testing result of damage zone(3#-15+250m)

## 3. Numerical simulation method for stress corrosion

To analyze the micro-mechanism of marble, the particle flow method is adopted, which is a discrete element method to simulate the movement and interaction of spherical particles.

### *3.1 Stress corrosion model*

The mechanical behaviors of rocks are controlled by their microscopic structures, and the rupture processes are caused by the reorganization of microscopic structures. Based on the linear elastic fracture mechanics model (LEFM), Potyondy and Cundall [12] introduced the damage rate into the Bonded-Particle Mode of PFC, which can simulate the stress corrosion behaviors of rock. In the LEFM model, the kinetics of crack growth are expressed as:

$$V=\begin{cases}0, & K_{\mathrm{I}}<K_{\mathrm{th}}\\ \alpha_1 \mathrm{e}^{\alpha_2(K_{\mathrm{I}}/K_{\mathrm{c}})}, & K_{\mathrm{th}}\le K_{\mathrm{I}}<K_{\mathrm{c}}\\ \infty, & K_{\mathrm{I}}\ge K_{\mathrm{c}}\end{cases} \tag{1}$$

where $K_c$ is the fracture toughness, $K_{th}$ is the stress-corrosion threshold, $\alpha_1$ and $\alpha_2$ are material constants.

In the stress corrosion (PSC) model, the rock is simplified as a cemented particle assemble, and the initiation and growth of cracks can be simulated by the contraction of the cement size. The kinetics of crack growth in the PSC model can be expressed as:

$$V=\begin{cases}0, & \bar{\sigma}<\bar{\sigma}_{\mathrm{a}}\\ \beta_1 \mathrm{e}^{\beta_2(\bar{\sigma}/\bar{\sigma}_{\mathrm{c}})}, & \bar{\sigma}_{\mathrm{a}}\le\bar{\sigma}<\bar{\sigma}_{\mathrm{c}}\\ \infty, & \bar{\sigma}\ge\bar{\sigma}_{\mathrm{c}}\end{cases} \tag{2}$$

where $\bar{\sigma}_c$ is the tensile strength of cement, $\bar{\sigma}_a$ is the micro-activation stress, and $\beta_1$ and $\beta_2$ are material constants.

When the driving stress exceeds the micro-activation stress, the growth rate of the crack can be computed by Eq. (2). Therefore, the stress corrosion of rock can be simulated by a PFC model.

*3.2 Validation of numerical model*

A rectangular particle assembly is created by PFC2D to simulate the Jinping marble, as shown in Figure 4.As rock is inhomogeneous, the contact-bond parameters between particles and particle size were normal distributed.

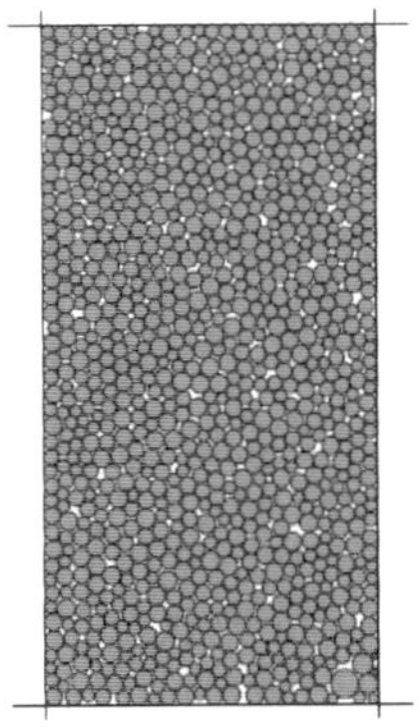

**Figure 4.** Assembled PFC rock models

The dimensions of the particle assembly are 31.7 mm wide by 63.4 mm high. Its parameters are listed in Table 1. According to the numerical experiment of uniaxial compression, the macro-mechanical parameters of the particle assembly are listed in Table 2, which are quite close to the experimental parameters of Jinping marble.

**Table 1.** Model micro-properties of rock

| Friction coefficient | Contact stiffness (GN/m) | Ratio of normal contact stiffness to shear contact stiffness | Normal contact-bond Strength (MPa) | Shear contact-bond Strength (MPa) |
|---|---|---|---|---|
| 1.0 | 40 | 2.5 | 70±15 | 70±15 |

**Table 2.** Simulated macro-mechanical parameters of rock

| Uniaxial compression strength (MPa) | Young's modulus (GPa) | Poisson's ratio | density (kg/m3) |
|---|---|---|---|
| 97.7 | 49.0 | 0.23 | 2600 |

A series of numerical biaxial compression stress corrosion experiments were conducted. The axial pressure $\sigma=0.4\sigma_c$ ~ $0.95\sigma_c$ ($\sigma_c$ is the uniaxial compression strength). The best obtained parameters of PSC model are $\beta_1=5\times10^{-14}$, $\beta_2=21$, $\overline{\sigma}_a=25$MPa. The simulated stress corrosion lifetime $t_f$ is illustrated in Figure 5. It indicates that the parameters are reasonable.

Most of the surrounding rocks are generally in triaxial stress state in tunnels. In order to validate the parameters of numerical specimens, the triaxial compression results of numerical and actual tests are compared, as shown in Figure 6. The triaxial compression strengths of numerical specimens and actual specimens are approximately consistent, which indicates that parameters are rational.

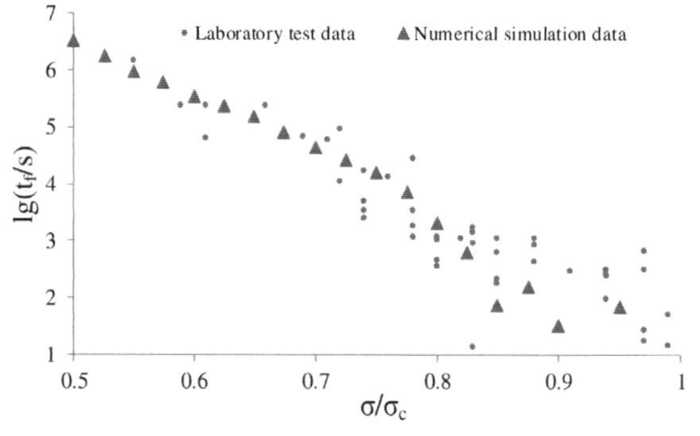


**Figure 5.** Comparison between numerical results and experimental results

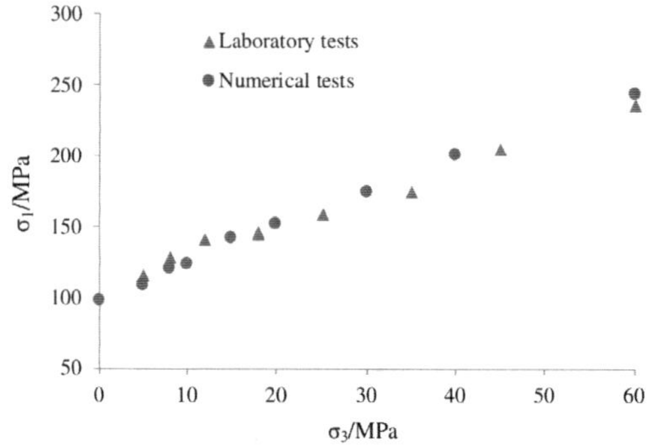


**Figure 6.** Comparison between numerical results and experimental data

## 4. Simulation of stress corrosion in tunnel rock mass

As illustrated in Figure 1, when the surrounding rocks of the tunnels were exposed a few hours or several months after excavation, they fractured gradually. However, their stress hadn't exceeded the strength. The stress state of the rocks on the tunnel walls was similar with the biaxial and triaxial compression experiment. Therefore, their stress corrosion rupture process is simulated in stress corrosion rupture zone and stress corrosion damage zone as illustrated in Figure 7.

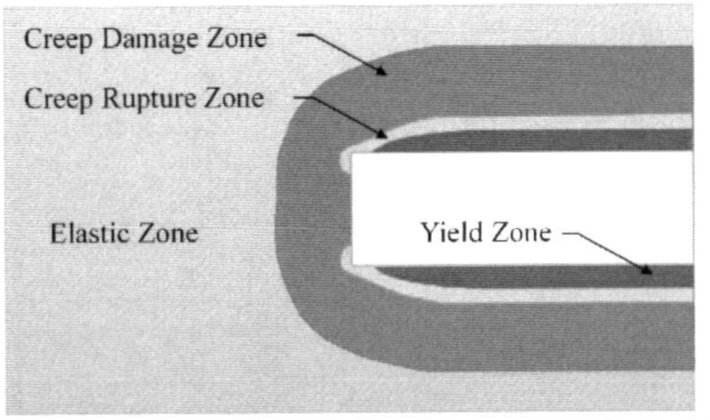


**Figure 7.** Yield zone and creep damage zone of tunnels

In the simulation, all the numerical specimens are the same. In biaxial stress corrosion simulation, the axial pressure keeps $0.4\sigma_c$~$0.95\sigma_c$. And the driving-stress ratio $\eta$ is defined as:

$$\eta=(\sigma_1-p_c)/(\sigma_p-p_c) \tag{3}$$

where $\sigma_P$ is the triaxial compression strength, $p_c$ is the confining stress.

180 cases are computed with $p_c$ are set to 5MPa, 8MPa, 10MPa, 15MPa, 20MPa, 30MPa, 40MPa and 60MPa respectively and $\eta$=0.4~0.95.

### *4.1. Rupture process on the tunnel walls*

*4.1.1 Rupture process and mechanism.* As illustrated in Figure 8, in the nucleation stage of microcracks, the vertical parallel-bond forces between particles are compressive (black line), and the horizontal parallel-bond forces are almost tensile (red line). Meanwhile, the forces between particles are not uniform because of the inhomogeneity of bonds. Thus, some axial tension microcracks begin nucleating at the weaker bonds firstly.

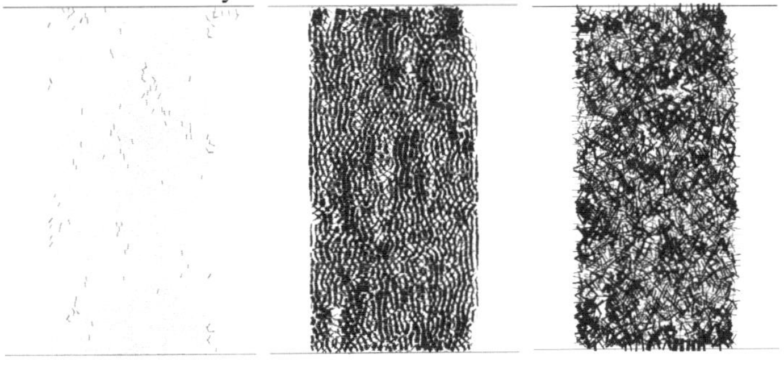

(a) (b) (c)

**Figure 8.** The microcracks and bond forces distribution in numerical models($t$=44000s, $\sigma$=0.7$\sigma_c$ , $p_c$=0), (a)microcracks,(b)normal bond forces,(c)shear bond forces

As illustrated in Figure 9, in the stable propagation stage of microcracks, the changes of bond forces are slow and not obvious, however, the nucleation of new microcracks causes some adjustments of the neighbor bond forces. Hence, some new microcracks nucleate near the old microcracks in this stage, and some nucleates at some new position.

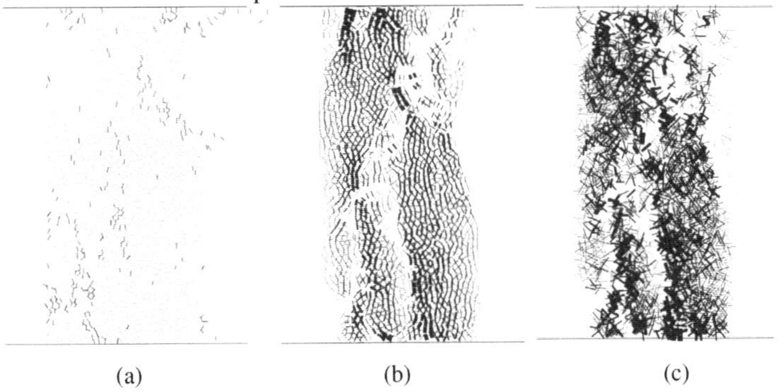

(a) (b) (c)

**Figure 9.** The microcracks and bond forces distribution in numerical models($t$=46400s, $\sigma$=0.7$\sigma_c$ $p_c$=0), (a)microcracks,(b)normal bond forces,(c)shear bond forces

As illustrated in Figure **10**, in the unstable propagation stage of microcracks, the maximum compressive, tensile and shear parallel-bond forces increase rapidly, so that a big number of microcracks are generated in some zone concentratedly. Since the rupture surface is rough, the shear forces on its both sides are high, then some shear microcracks appears.

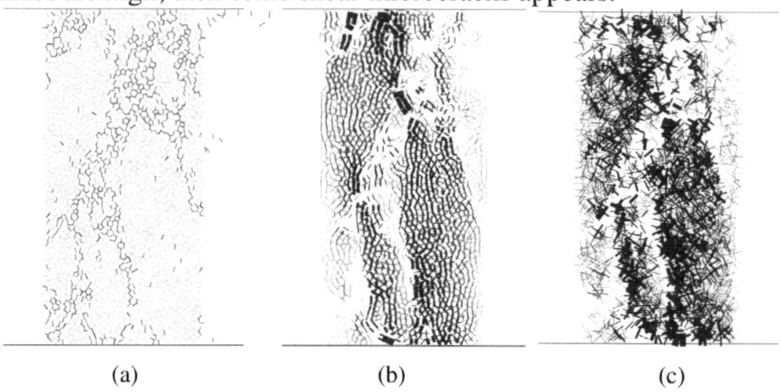


**Figure 10.** The microcracks and bond forces distribution in numerical models($t$=46500s, $\sigma$=0.7$\sigma_c$ , $p_c$=0), (a)microcracks,(b)normal bond forces,(c)shear bond forces

*4.1.2 Influence of time and stress.* As shown in Figure 11, the stress corrosion lifetime $t_f$ of the specimen is an infinite number if the load $\sigma_1$ is less than about 0.425$\sigma_c$, which means it will not rupture finally. The history of microcrack number under different axial stress is plotted in Figure 12. It shows that some microcracks have nucleated before the beginning of stress corrosion tests. Then, in the propagation of microcracks, it shows that the stress corrosion rupture process can be divided into 4 stages, and which is like the brittle failure process:

(a) when $t \leq 0.1\ t_f$, the number of microcracks increases rapidly, but the growth rates decrease gradually. And the growth rates under different loads are similar to each other in this stage.

(b) When $0.1\ t_f < t \leq 0.6\ t_f$, the growth rate of microcracks is lower, and the number is increasing linearly with $t/t_f$.

(c) When $0.6\ t_f < t \leq t_f$, the growth rate of microcracks increases exponentially with $t/t_f$, which means the specimen ruptures acceleratively. In this stage, the lower the load is, the larger the growth rate becomes.

(d) When $t = t_f$, the growth rate of microcracks leaps in a little time.

In the aspect of rupture pattern, the appearance of the damaged specimens is slightly different under different stress. The microcracks are more concentrated in the defective zone when the axial stress is lower (Figure 13), which is similar with the actual results (Figure 14).

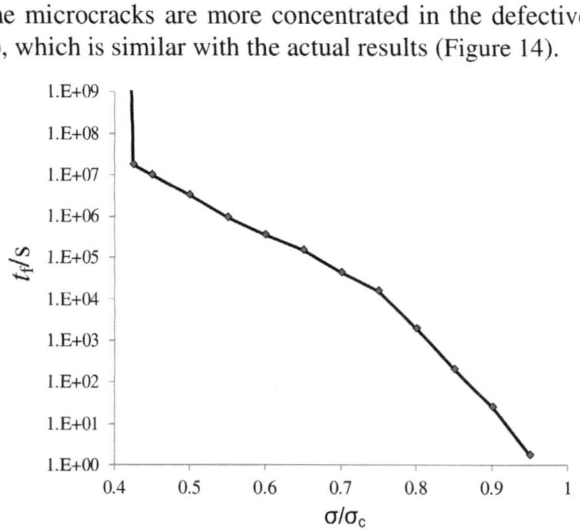


**Figure 11.** The relationship between stress corrosion lifetime $t_f$ and stress when $p_c$=0

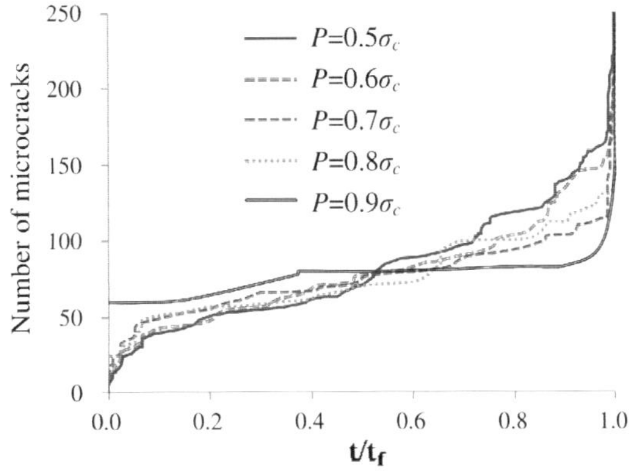


**Figure 12.** The relationship between the number of microcracks and the time when $p_c$=0

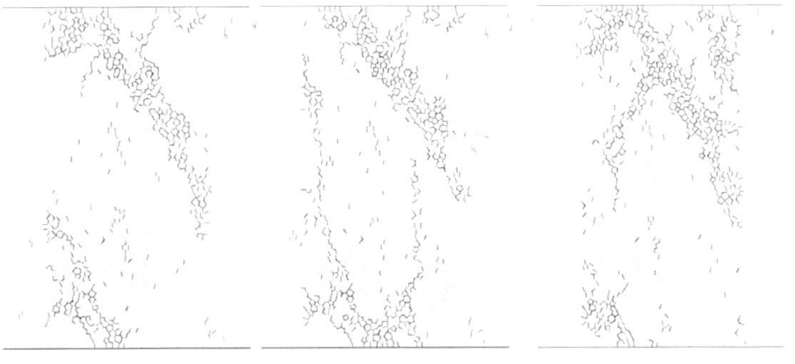

(a) (b) (c)

**Figure 13.** Rupture patterns of numerical samples when $p_c$=0 (a) $\sigma$=0.5$\sigma_c$, (b) $\sigma$=0.6$\sigma_c$, (c) $\sigma$=0.8$\sigma_c$

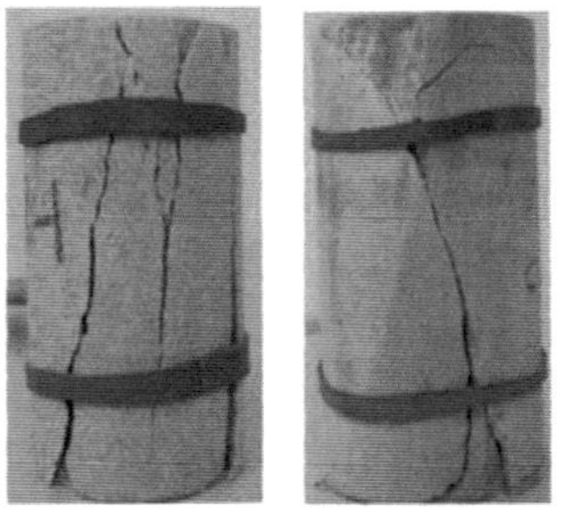

(a) (b)

**Figure 14.** Rupture patterns of actual experiment samples when $p_c$=0 (Liu, 2011): (a) $\sigma$=0.61$\sigma_c$, (b) $\sigma$=0.97$\sigma_c$

*4.2. Rupture process near the tunnel walls*

Near the tunnel walls, the confining stress of the surrounding rock is low, and driving-stress ratio $\eta$ may be high, which are influenced by the ratio of vertical and horizontal geostress. In this case, the rupture processes of the specimens are simulated under the condition of $p_c$=5MPa.

In the case of the driving-stress ratios $\eta$ are low, the simulation results of an example are shown in Figure 15, in which $\eta$=0.55. In the microcrack growth stage, a lot of microcracks concentrate at the stress concentrated zone. Then the macro crack connects most of those microcracks. And it is found that when $\eta$ is less than 0.55, the specimens will just damage but not rupture.

If $\eta$=0.9, there are much less microcracks before rupture than that in the $\eta$=0.55 case. As the Figure 16 shows, the major crack is formed only in 1 second, and a large number of transverse microcracks nucleate. Compared with the biaxial compression tests, the major cracks are located on the diagonal of the specimens.

When the confining stress of the surrounding rock is low, the number of microcracks over the relative time $t/t_f$ is shown in Figure 17. If the driving-stress ratio $\eta$ is lower, the number of microcracks may increase gradually before rupture. And, if $\eta$ is higher, the number of microcracks may increase more rapidly in the rest of its life. If $\eta$ is close to 1, the number of microcracks leaps continually.

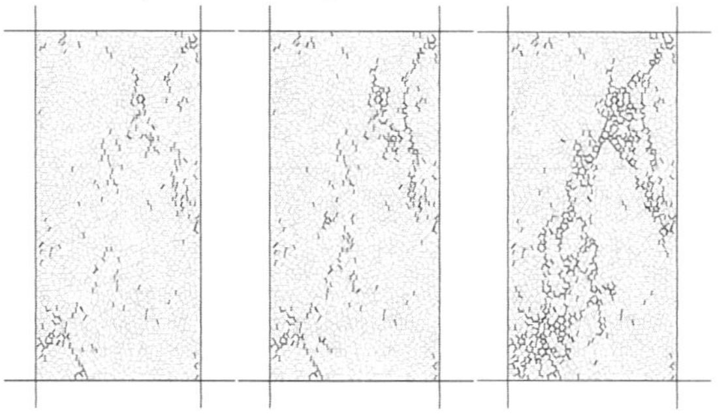

(a) (b) (c)

**Figure 15.** Creep rupture process when $p_c$=5MPa and $\eta$=0.55
(a) $t$=12300000s, (b) $t$= 12475000 s, (c) $t$=12476000s

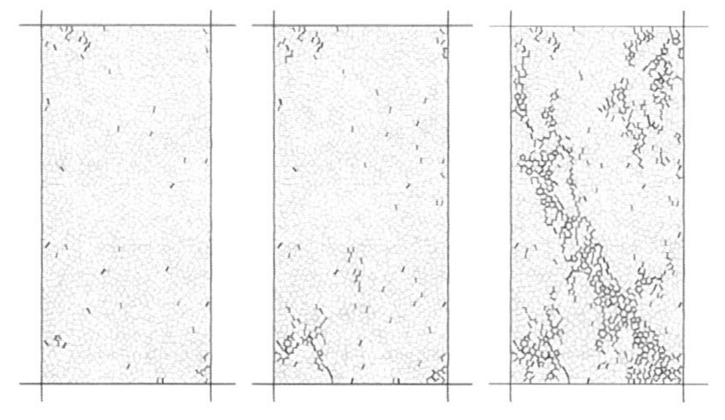

(a) (b) (c)

**Figure 16.** Creep rupture process when $p_c$=5MPa and $\eta$=0.9
(a) $t$=10s, (b) $t$= 195 s, (c) $t$=200s

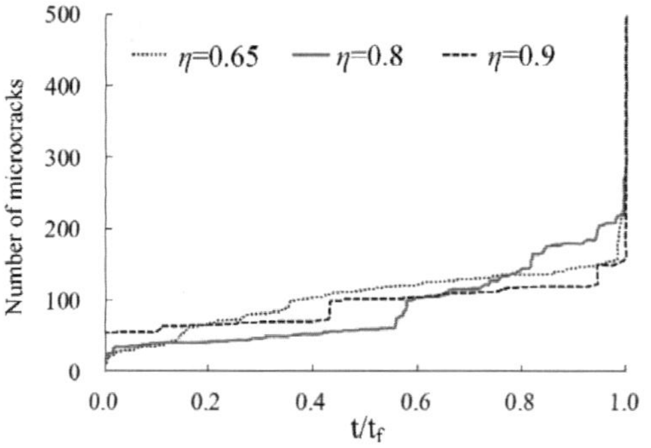


**Figure 17.** The relationship between the number of microcracks and the time when $p_c$=5MPa

### *4.3. Rupture process near the yield zone*

The rocks near the yield zone rupture more easily, whose confining stress $p_c$ is relatively high. The simulation results of $p_c$=15MPa are illustrated in Figure 18 and Figure 19. If the confining stress $p_c$ is high, more microcracks will nucleate before rupture. When $\eta$ is less than 0.65, the specimens will just damage but not rupture. When driving-stress ratio $\eta$=0.9, there are more shear microcracks than in $\eta$=0.65 case. And rupture occurs more rapidly, which is similar with the cases of $p_c$=5MPa.

As illustrated in Figure 20, if the confining stress $p_c$ is high, the number of microcracks increases steadily with time before rupture. the growth rate of microcracks in the former and latter part of their life-span is higher than in the middle part. The total number at the rupture time is more than that in the

$p_c$=5MPa case.

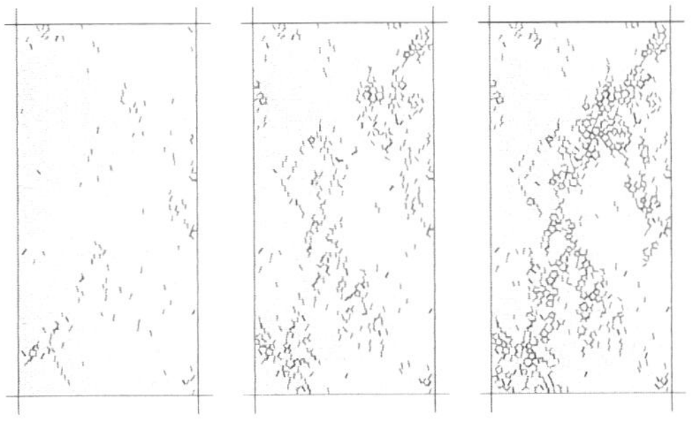

(a) (b) (c)

**Figure 18.** Creep rupture process when $p_c$=15MPa and $\eta$=0.65
(a) $t$=1000000s, (b) $t$= 5650000 s, (c) $t$=5700000s

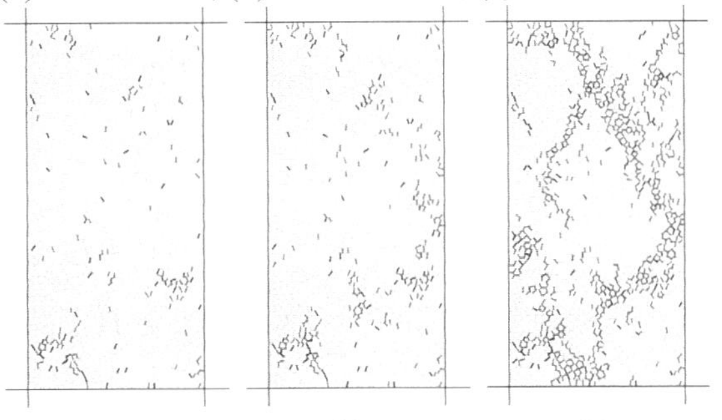

(a) (b) (c)

**Figure 19.** Creep rupture process when $p_c$=15MPa and $\eta$=0.9
(a) $t$=30s, (b) $t$= 325 s, (c) $t$=326s

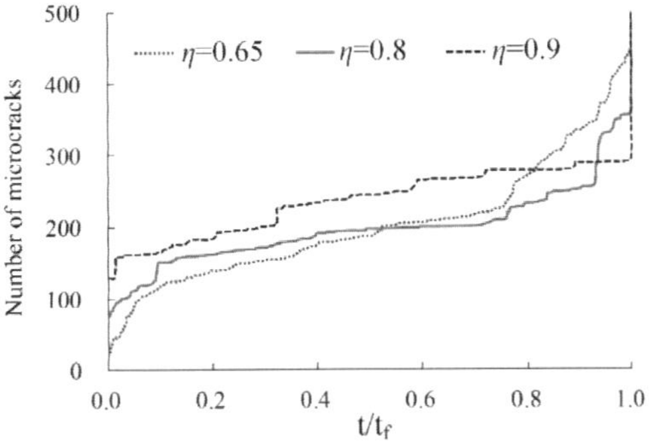


**Figure 20.** The relationship between the number of microcracks and the time when $p_c$=15MPa

## *4.4. Rupture process far from the tunnel walls*

Far from the tunnel walls, the confining stress $p_c$ of surrounding rock is close to the in-situ stress and driving-stress ratio is low. Therefore, the phenomenon of stress corrosion rupture rarely happens, but the rocks may be damaged gradually.

The simulation results in which $p_c$=40MPa are illustrated in Figure 21 and Figure 22. In the case of driving-stress ratio $\eta$=0.65, microcracks nucleate only in about 30 days, then the damage cannot develop further. When driving-stress ratio $\eta$=0.8, there are more shear microcracks in the specimen than that in $\eta$=0.65 case. And rupture occurs more rapidly, which is similar with the cases of $p_c$=5MPa. Furthermore, the amount of microcracks is also larger than other cases.

As shown in Figure 23, when the confining stress of surrounding rock is very high, the rock specimens are hard to be damaged. In the former stage of its life-span, the number of microcracks grows rapidly. When the specimen ruptures, the number of microcracks increases suddenly. Therefore, the rupture process is like the ductile failure process.

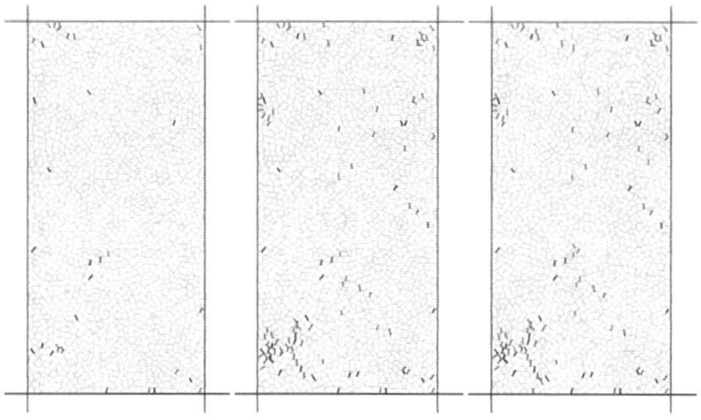

(a) (b) (c)

**Figure 21.** Creep rupture process when $p_c$=40MPa and $\eta$=0.65
(a) $t$=0, (b) $t$=35 days, (c) $t$=5 years

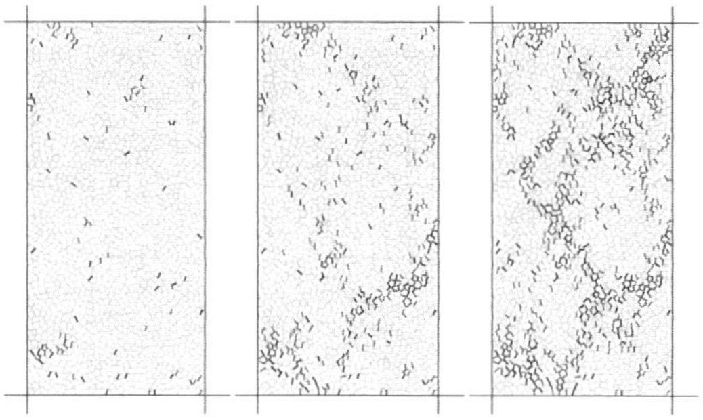

(a) (b) (c)

**Figure 22.** Creep rupture process when $p_c$=40MPa and $\eta$=0.8
(a) $t$=12 days, (b) $t$=36 days, (c) $t$=141 days

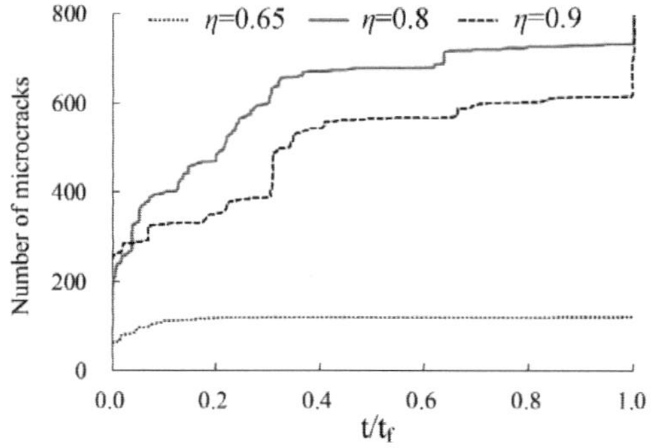


**Figure 23.** The relationship between the number of microcracks and the time when $p_c$=40MPa

### *4.5. Influence of stress on the stress corrosion lifetime*

180 cases are simulated to analyze the influence of stress state on the stress corrosion lifetime $t_f$, in which $p_c$ =5~60MPa and $\eta$=0.4~0.95. The results are shown in Figure 24. If the confining stress keeps constant, there is a linear relationship between the logarithm of stress corrosion lifetime lg$t_f$ and driving-stress ratio $\eta$. It indicates that $t_f$ will increase rapidly when $\eta$ becomes lower. On the other hand, $t_f$ will also increase when $p_c$ becomes higher. The simulation results show that there is a threshold of driving-stress ratio $\eta$ if the confining stress $p_c$ is a constant, which controls whether the rock specimens will rupture or not.

The higher the confining stress $p_c$, the stronger the influence of driving-stress ratio $\eta$ on stress corrosion lifetime $t_f$. However, if $\eta$>0.9, the stress corrosion lifetimes $t_f$ of all cases are close, which indicates that the damage and rupture process are unstable when loading close to their strength.

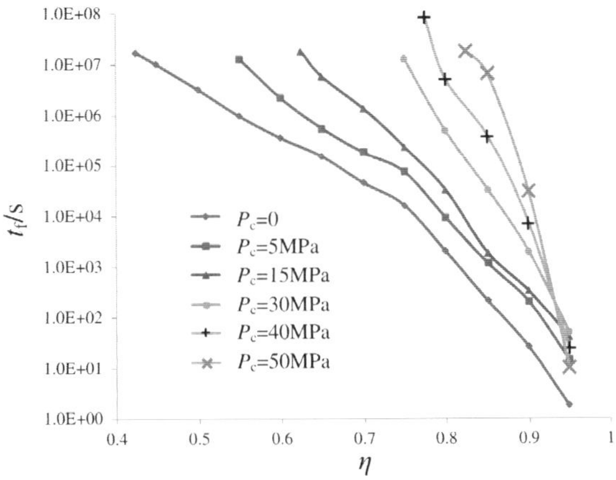


**Figure 24.** Relationship between $t_f$ and $\eta$ with different $p_c$

## 5. Prediction of the stress corrosion rupture zone of the tunnel

### *5.1. Long-term strength*

There is a critical stress state at which microcracks begin to nucleate [18]. This critical stress state can be regarded as the long-term strength of rock. In actual uniaxial compression experiments and Acoustic Emission experiment, the long-term strength of Jinping marble is about $\sigma_{cf}$=0.4$\sigma_c$~0.5$\sigma_c$ [18]. And the long-term strength of the numerical model is $\sigma_{cf}$=0.425$\sigma_c$ in this study, which is agreed with the laboratory result well. It indicates that the long-term strength can be obtained by numerical simulation. Therefore, the thresholds of axial stress $\sigma_{1f}$ with different confining stress $\sigma_3$ are obtained by numerical tests, as shown in Figure 25. the $\sigma_{1f}$ increases linearly with the confining stress $\sigma_3$. Therefore, the long-term strength of Jinping marble can be expressed as:

$$\sigma_{1f} = 1.9\sigma_3 + 0.49\sigma_c \tag{4}$$

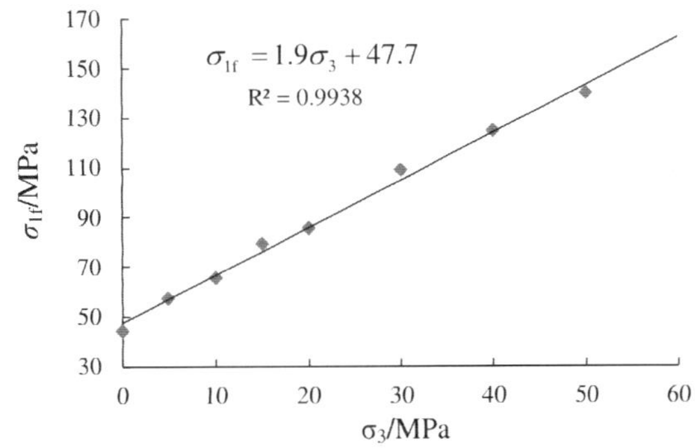


**Figure 25.** Relationship between creep rupture strength and confining stress

### *5.2. Prediction of stress corrosion rupture zone*

For the safety of the diversion tunnels excavated by TBM of Jinping II hydropower station, the range of transient excavated damage zone and stress corrosion rupture zone was predicted by finite element method. The depth of the tunnel section is 1770m, the diameter is 12.4m, the vertical geostress is $S_v$=55MPa, and the horizontal geostress is $S_h$=46MPa. The uniaxial compression strength of the intact rock is $\sigma_{ci}$=100MPa. The Hoek-Brown strength parameters of the rock mass $m_b$=3.08, $s$=0.0357. The elastic modulus of rock mass is $E$=38.83GPa, and Poisson's ratio $\mu$=0.2. The yield and stress corrosion rupture zone of the diversion tunnels are shown in Figure 26 and Figure 27, In which the stress state of rock satisfies $\sigma_1 \geq 1.9\sigma_3 + 47.7$.

The maximum thickness of the yield rock mass is about 1.63m, and the maximum thickness of the stress corrosion rupture zone exceeds 1.4m if no support is installed. However, the bolts and liners change the stress state of the surrounding rock mass, and they reduce the thickness of the yield zone to 0.91m, meanwhile, the final thickness of the rupture zone will be about 2.4m. In fact, the actual rupture range should be slightly larger than that of the simulation, because the strength of the yield rock will decrease and the bolts are installed several hours or days later.

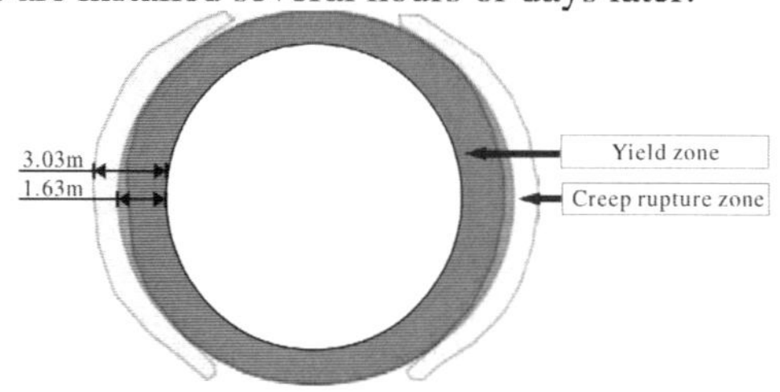


**Figure 26.** Yield zone and creep rupture zone of the tunnel without supports

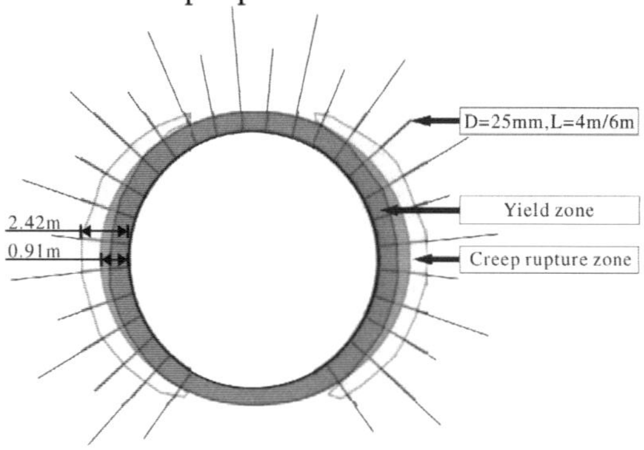


**Figure 27.** Yield zone and creep rupture zone of the tunnel installing bolts and liners

## 6. Conclusions

Spalling and rockburst are the serious geology problems in the diversion tunnels of Jinping II Hydropower Station project. To investigate the stress corrosion behaviors of the diversion tunnels, numerical simulations verified by experimental tests are performed, and the following conclusions are drawn:

(1) According to the development history of microcracks, the stress corrosion rupture process of surrounding rock can be divided into 4 stages, i.e., rapid growth, slow growth, rapid growth and instant leap. The critical points of the 4 stages are 0, $0.1t_f$, $0.6t_f$, $t_f$.

(2) On or near the tunnel wall, most stress corrosion microcracks are induced by tensile stress and parallel to the direction of the maximum principle stress. And the stress corrosion rupture process is like a brittle failure process. In the rocks near the yield zone, the amount of microcracks is larger at the rupture stage than that on the tunnel wall, and the ratio of shear microcracks becomes larger with increasing confining stress, which makes the stress corrosion rupture process look like a ductile failure process.

(3) There is a linear relationship between the logarithm of stress corrosion lifetime $\lg t_f$ and driving-stress ratio $\eta$ when the confining stress keeps constant. Stress corrosion lifetime $t_f$ will increase when $p_c$ becomes higher or driving-stress ratio becomes lower.

## 7. Data Availability

The data used to support the findings of this study are available from the corresponding author upon request.

## 8. Conflicts of Interest

The authors declare that there is no conflict of interest regarding the publication of this paper.

## Acknowledgments

The work reported in this paper is financially supported by the Key Research and Development

Project of Hubei Province, China (No. 2020BCA084), the Natural Science Foundation of Hubei Province, China (Nos. 2020CFA043 and 2020CFB428).

## 9. References

[1] Cao S, Xian X. Study on the characteristics of creep and damage of coal and other rocks. *Chinese Journal of Rock Mechanics Engineering* 2001; 20: 817-822.

[2] Chen W, Zhu W, Li S. Coupled rheological model study on fracture-damage problem for jointed rocks and its applications. *JHYDRAUL ENG-ASCE* 1999; 30: 33-37.

[3] Chen Y. Test study on creep fracture characteristics of hard rocks. *ACTA MECH SINICA* 2003; 35: 480-483.

[4] Holcomb DJ, Martin RJ. Determining peak stress history using acoustic emissions, in:Ashworth, E., Balkema, A.A. (Eds.), 26th U.S. Symposium on Rock Mechanics. *Rapid City, Rotterdam*, pp, 1985: 715-722.

[5] Kang DH, Yun TS, Lau YM, Wang YH. Simulation of soil creep and associated evolution of pore characteristics using discrete element method and image processing. *Comput. Geotech* 2012; 39: 98–106.

[6] Kwok CY, Bolton MD. DEM simulations of thermally activated creep in soils. *GEOTECHNIQUE* 2010; 60: 425–433.

[7] Ling J. Study on the mesoscopical characteristics of rock damage under compressive loading. *Journal of Tongji University* 1993; 21: 219-226.

[8] Liu N. Time-dependent failure of deep buried marble and estimation on long-term stability of tunnel. *Power China Huadong Engineering Corporation Limited*, Hangzhou; 2011.

[9] Liu N, Zhang C, Chu W. Simulating time-dependent failure of deep marble with particle flow code. *Chinese Journal of Rock Mechanics Engineering* 2011; 30: 1989-1997.

[10] Martin CD, Chandler NA. The progressive fracture of Lac du Bonnet granite. *INT J ROCK MECH MIN* 1994; 31: 643-659.

[11] Potyondy DO. Simulating stress corrosion with a bonded-particle model for rock. *INTJROCKMECHMIN* 2007; 44: 677-691.

[12] Potyondy DO, Cundall PA. A bonded-particle model for rock. INT J ROCK MECH MIN 2004; 41: 1329-1364.

[13] Sun J. Rheological mechanics and its advance in engineering applications. *Chinese Journal of Rock Mechanics Engineering* 2007; 26: 1081-1106.

[14] Sun J, Chen M, Chen B, Lu W, Zhou C. Study of mechanical behavior of deep-buried marble at Jinping II hydropower station. *Chinese Journal of Rock Mechanics and Engineering* 2011; 32: 1891-1898.

[15] Xiao H, Zhou W, Yang R. A stability analysis of rheological damage of high rock slope of the permanent ship-lock, the Three Gorges Project in China. *China Civil Engineering Journal* 2000; 33: 94-98.

[16] Xie H. Damage mechanics of rock and concrete. *Press of Chinese University of Mining Science and Technology*, Xuzhou; 1990.

[17] Yang C, Chen F, Zeng Y. A study on creep damage of rock salt. *Chinese Journal of Rock Mechanics Engineering* 2002; 21: 1602-1604.

[18] Zhang C, Chen X, Hou J, Chu W. Study of mechanical behavior of deep-buried marble at Jinping II hydropower station. *Chinese Journal of Rock Mechanics and Engineering* 2010; 29: 1999-2010.

[19] Zhang Q, Yang W, Zhang J, Yang C. Variable parameters-based creep damage constitutive model and engineering application. *Chinese Journal of Rock Mechanics Engineering* 2009; 28: 732-740.

[20] Zhao B. Study on creep and damage law of deep rock. *Xihua University*, Hefei; 2008.

# AUTHOR INDEX

**Institute of Physics**
Dirac House, Temple Back
Bristol BS1 6BE UK

ISSN: 1757-8981
ISBN 978-1-7138-3430-4

9 781713 834304

CURRAN ASSOCIATES INC.
proceedings.com